Series on Analysis, Applications and Computation – Vol. 2

Complex Analysis

Series on Analysis, Applications and Computation

Published

Vol. 1: Boundary Values and Convolution in Ultradistribution Spaces
by R D Carmichael, A Kamiński & S Pilipović

Vol. 2: Complex Analysis
by M W Wong

Series on Analysis, Applications and Computation – Vol. 2

Complex Analysis

M W Wong

York University, Canada

NEW JERSEY • LONDON • SINGAPORE • BEIJING • SHANGHAI • HONG KONG • TAIPEI • CHENNAI

Published by

World Scientific Publishing Co. Pte. Ltd.
5 Toh Tuck Link, Singapore 596224
USA office: 27 Warren Street, Suite 401-402, Hackensack, NJ 07601
UK office: 57 Shelton Street, Covent Garden, London WC2H 9HE

British Library Cataloguing-in-Publication Data
A catalogue record for this book is available from the British Library.

COMPLEX ANALYSIS
Series on Analysis, Applications and Computation — Vol. 2

ISBN-13 978-981-281-107-3
ISBN-10 981-281-107-9

Printed in Singapore.

Preface

This book is the culmination of teaching complex analysis to many groups of students at York University for many years. It is intended for a one-semester course for advanced undergraduate students and also first-year graduate students in mathematics. In a nutshell, it is a straightforward and coherent account of a body of knowledge in complex analysis from complex numbers to Cauchy's integral theorems and formulas to more advanced topics such as automorphism groups, the Schwarz problem in partial differential equations and boundary behavior of harmonic functions. While a few rudimentary facts in modern algebra are used in Chapters 21 and 22, a basic undergraduate course in real analysis is the only prerequisite for a complete understanding of the book.

It is clear from the table of contents that the topics are standard and can be found in different existing textbooks and monographs. The subject matter on complex analysis, especially on one-variable complex analysis, is so mature that in almost any textbook of this kind, the novelty has to be sought in the choice of topics, the genesis of the presentation and the lucidity of the exposition. The predilection for topics in this book is dictated by personal experiences and the inevitably personal view on the relevance to the forefront of research in analysis and partial differential equations. Nothwithstanding the importance of the geometric and topological nature of complex analysis to many experts, the emphasis of this book is decisively on analysis. While this is a book on mathematics with no explicit mention of applications to other sciences or engineering, it is evident that the bulk of the book consisting of Chapters 1–18 contains the bread and butter for students in pure and applied mathematics using complex analysis as a mathematical tool or learning the subject for its own sake.

To render the book accessible and elementary and to make sure that a one-semester course can convey enough topics of sufficient depth to the students, some care has to be taken in the treatment of Cauchy's integral theorem. In this book Cauchy's integral theorem is stated precisely in sufficient generality, but a proof of it is only given for a rectangle in a simply connected domain. Exercises for each chapter are included. Some of them contain ramifications of the results presented in detail in the text. The amount is minimal and the temptation of including a lot of drill exercises is avoided. Students should do all of them.

The first eighteen chapters constitute a good first course in complex analysis for undergraduate students in mathematics, physical sciences and engineering. A course tailored for more advanced undergraduate students and first-year graduate students may be based on Chapters 5–23. An ideal course is to study the book in detail from cover to cover in one semester.

It is hardly necessary to give an extensive list of references on complex analysis at the level of this book. Excellent books on the subject abound and can be found easily in any university library. Collected in the bibliography are books and papers on complex analysis that have been useful for the writing of this book. Useful references on real analysis and modern algebra used in this book are provided. Also included in the bibliography are references that are extensions of the more advanced topics in the last three chapters of the book. The value of demonstrating to the students the relevance of such a well-established and classic subject as complex analysis to mainstream mathematics and some research topics of current interest is enormous.

M. W. Wong

Contents

Chapter 1

Complex Numbers

Let $\mathbb{R}$ be the set of all real numbers. Then a complex number is of the form $a + ib$, where a and b are in $\mathbb{R}$ and $i^2 = -1$. We denote the set of all complex numbers by $\mathbb{C}$.

Definition 1.1. Let $a + ib$ and $c + id$ be complex numbers. Then we say that $a + ib = c + id$ if and only if $a = c$ and $b = d$.

Remark 1.2. Using the formula $i^2 = -1$, complex numbers are added, subtracted, multiplied and divided like real numbers.

Example 1.3. Compute $(2 + i3)(3 - i4)$.

Solution $(2 + i3)(3 - i4) = 6 - i8 + i9 - 12i^2 = 6 + i + 12 = 18 + i$.

Example 1.4. Compute $\frac{2+i3}{1+i2}$.

Solution We rationalize the denominator and we get

$$\frac{2 + i3}{1 + i2} = \frac{(2 + i3)(1 - i2)}{(1 + i2)(1 - i2)} = \frac{2 - i4 + i3 - 6i^2}{1 - 4i^2} = \frac{8 - i}{5} = \frac{8}{5} - i\frac{1}{5}.$$

Let $z = a + ib$ be a complex number. Then we call a the real part of z and b the imaginary part of z. We sometimes write $a = \operatorname{Re} z$ and $b = \operatorname{Im} z$. The complex conjugate $\overline{z}$ of z is defined by

$$\overline{z} = a - ib.$$

The absolute value $|z|$ of z is defined by

$$|z| = \sqrt{a^2 + b^2}.$$

We can now give a geometric interpretation of complex numbers. We identify $\mathbb{C}$ as the xy-plane $\mathbb{R}^2$ and we identify the complex number $z = a+ib$

as the point (a, b) in $\mathbb{R}^2$. Then $\overline{z}$ is the point $(a, -b)$, which is the mirror image of z with respect to the x-axis. $|z|$ is then simply the distance between z and the origin.

Theorem 1.5. *Let z be a complex number. Then*

(1) $|z|^2 = z\overline{z}$,
(2) $\operatorname{Re} z = \frac{z+\overline{z}}{2}$,
(3) $\operatorname{Im} z = \frac{z-\overline{z}}{2i}$,
(4) $\overline{\overline{z}} = z$,
(5) $|\overline{z}| = |z|$.

Proof Let $z = a + ib$. Then

$$z\overline{z} = (a+ib)(a-ib) = a^2 - b^2i^2 = a^2 + b^2 = |z|^2.$$

This is Part (1). For Parts (2) and (3), we note that

$$\frac{z+\overline{z}}{2} = \frac{a+ib+a-ib}{2} = a = \operatorname{Re} z$$

and

$$\frac{z-\overline{z}}{2i} = \frac{a+ib-a+ib}{2i} = b = \operatorname{Im} z.$$

The last two parts can be seen most easily from the geometric interpretation of complex numbers. □

Theorem 1.6. *Let z_1 and z_2 be complex numbers. Then*

(1) $\overline{z_1 \pm z_2} = \overline{z_1} \pm \overline{z_2}$,
(2) $\overline{z_1 z_2} = \overline{z_1}\,\overline{z_2}$,
(3) $\overline{\left(\frac{z_1}{z_2}\right)} = \frac{\overline{z_1}}{\overline{z_2}}$.

The proof of Theorem 1.6 is easy and hence omitted.

Exercises

(1) Prove that for all complex numbers z,

$$\operatorname{Re}(iz) = -\operatorname{Im} z.$$

(2) Prove that if z is a complex number such that $\operatorname{Im} z > 0$, then

$$\operatorname{Im}\left(\frac{1}{z}\right) < 0.$$

(3) Prove that if z and w are complex numbers such that $z + w$ is a real number and zw is a negative real number, then z and w are both real numbers.

(4) Prove that if z is a complex number such that $|z| = \operatorname{Re} z$, then z is a nonnegative real number.

(5) Draw the set of all the points z in $\mathbb{C}$ satisfying each of the following relations.

(a) $|2z - i| = 4$
(b) $|z| = \operatorname{Re} z + 2$
(c) $|3z + i| < 2$
(d) $|z - 1| + |z + 1| = 7$.

Chapter 2

Arguments and Polar Forms of Complex Numbers

The notion of a phase or an argument of a complex number is what makes complex numbers have a flavor different from real numbers. To see what it is, let $z = x + iy$ be a nonzero complex number. Then in terms of polar coordinates, z can be identified as (r, θ), where

$$r = |z| = \sqrt{x^2 + y^2}$$

and

$$\begin{cases} x = r\cos\theta, \\ y = r\sin\theta. \end{cases}$$

Note that if $z = (r, \theta)$, then $z = (r, \theta + 2k\pi)$, where k is any integer. We call $\theta + 2k\pi$ an argument of z and denote it by $\arg z$. Thus,

$$\arg z = \theta + 2k\pi, \quad k = 0, \pm 1, \pm 2, \ldots.$$

So, if we are able to find one value or branch for $\arg z$, then all the values or branches for $\arg z$ can be obtained by adding $2k\pi$ to it. Thus, an argument is reminiscent of an antiderivative in calculus in the sense that if one antiderivative can be found, then all antiderivatives can be found by adding a constant C to it.

With these notions in place, we can express a nonzero complex number $z = x + iy$ as

$$z = r(\cos\theta + i\sin\theta),$$

which is known as the polar form of z.

Example 2.1. Find $\arg(1 + i\sqrt{3})$ and express $1 + i\sqrt{3}$ in polar form.

Solution Let $z = 1 + i\sqrt{3}$. Then

$$r = |z| = \sqrt{1+3} = \sqrt{4} = 2,$$

and

$$\arg(1+i\sqrt{3}) = \frac{\pi}{3} + 2k\pi, \quad k = 0, \pm1, \pm2, \ldots.$$

Thus, a polar form is given by

$$z = 1 + i\sqrt{3} = 2\left(\cos\frac{\pi}{3} + i\sin\frac{\pi}{3}\right).$$

An advantage of using the polar form is that it provides geometric insight into the multiplication of complex numbers.

Example 2.2. Let

$$\begin{cases} z_1 = r_1(\cos\theta_1 + i\sin\theta_1), \\ z_2 = r_2(\cos\theta_2 + i\sin\theta_2). \end{cases}$$

Find a polar form for $z_1 z_2$.

Solution

$$\begin{aligned} z_1 z_2 &= r_1 r_2(\cos\theta_1 + i\sin\theta_1)(\cos\theta_2 + i\sin\theta_2) \\ &= r_1 r_2\{(\cos\theta_1\cos\theta_2 - \sin\theta_1\sin\theta_2) + i(\sin\theta_1\cos\theta_2 + \cos\theta_1\sin\theta_2)\} \\ &= r_1 r_2\{\cos(\theta_1+\theta_2) + i\sin(\theta_1+\theta_2)\}. \end{aligned}$$

Remark 2.3. From Example 2.2, we see that for all complex numbers z_1 and z_2,

$$|z_1 z_2| = |z_1||z_2|.$$

Furthermore, if z_1 and z_2 are nonzero, then

$$\arg(z_1 z_2) = \arg z_1 + \arg z_2.$$

Example 2.4. Let z_1 and z_2 be as in Example 2.2. Find $\frac{z_1}{z_2}$ in polar form.

Solution Since division is the inverse of multiplication, we get

$$\frac{z_1}{z_2} = \frac{r_1}{r_2}\{\cos(\theta_1-\theta_2) + i\sin(\theta_1-\theta_2)\}.$$

Remark 2.5. From Example 2.4, we get

$$\arg\frac{z_1}{z_2} = \arg z_1 - \arg z_2.$$

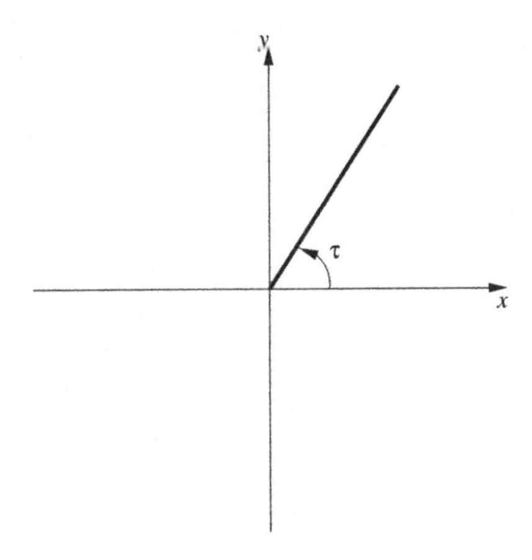

Fig. 2.1 A Cut Plane

We end this chapter with an important notion related to the argument of a complex number. For every nonzero complex number z, if θ is a value or branch of $\arg z$, then

$$\arg z = \theta + 2k\pi, \quad k = 0, \pm 1, \pm 2, \ldots.$$

Let τ be a real number. The plane $\mathbb{C}_\tau$ given by

$$\mathbb{C}_\tau = \{(r, \theta) : r > 0,\ \tau < \theta \le \tau + 2\pi\}$$

is known as the cut plane along the branch cut $\{(r, \tau) : r > 0\}$. The branch of $\arg z$ that lies in $(\tau, \tau + 2\pi]$ is denoted by $\arg_\tau z$. Of particular importance is the principal branch of $\arg z$, which we take to be the branch in $(-\pi, \pi]$. The principal branch of $\arg z$ is therefore the branch $\arg_{-\pi} z$ and is also denoted by $\operatorname{Arg} z$.

Example 2.6. Let $z = 1 + i$. Find $\operatorname{Arg} z$ and $\arg_{\frac{\pi}{2}} z$.

Solution Geometrically, we see that

$$\arg z = \frac{\pi}{4} + 2k\pi, \quad k = 0, \pm 1, \pm 2, \ldots.$$

Thus, $\operatorname{Arg} z = \frac{\pi}{4}$ and $\arg_{\frac{\pi}{2}} z = \frac{3\pi}{4}$.

Exercises

(1) Find the argument for each of the following complex numbers.

(a) $-3 + i3$

(b) $(1-i)(-\sqrt{3}+i)$
(c) $\frac{-1+i\sqrt{3}}{2+i2}$.

(2) Find the principal argument for each of the complex numbers in the preceding exercise.

Chapter 3

Exponentials, Powers and Roots

We all know what e^θ is if θ is a real number. What is $e^{i\theta}$? To answer this question, we again use the rule that complex numbers are treated as real numbers in calculations. If we use the power series for $e^{i\theta}$, then we get

$$\begin{aligned} e^{i\theta} &= \sum_{n=0}^{\infty} \frac{(i\theta)^n}{n!} \\ &= 1 + \frac{i\theta}{1!} + \frac{(i\theta)^2}{2!} + \frac{(i\theta)^3}{3!} + \cdots \\ &= \left(1 + \frac{(i\theta)^2}{2!} + \frac{(i\theta)^4}{4!} + \cdots\right) + \left(\frac{i\theta}{1!} + \frac{(i\theta)^3}{3!} + \frac{(i\theta)^5}{5!} + \cdots\right) \\ &= \left(1 - \frac{\theta^2}{2!} + \frac{\theta^4}{4!} - \cdots\right) + i\left(\theta - \frac{\theta^3}{3!} + \frac{\theta^5}{5!} - \cdots\right) \\ &= \cos\theta + i\sin\theta. \end{aligned}$$

Hence we have the celebrated Euler's identity to the effect that

$$e^{i\theta} = \cos\theta + i\sin\theta$$

for all real numbers θ. Another way to obtain Euler's identity is given in Chapter 6.

Recall that the polar form of a nonzero complex number z is given by

$$z = r(\cos\theta + i\sin\theta).$$

Thus, using Euler's identity, the polar form becomes

$$z = re^{i\theta}.$$

Example 3.1. Compute $e^{\pi i}$.

Solution $e^{\pi i} = \cos\pi + i\sin\pi = -1$.

Example 3.2. Prove that for every positive integer n,

$$(\cos\theta + i\sin\theta)^n = \cos n\theta + i\sin n\theta.$$

This formula is known as De Moivre's formula.

Solution

$$(\cos\theta + i\sin\theta)^n = (e^{i\theta})^n = e^{in\theta} = \cos n\theta + i\sin n\theta.$$

Example 3.3. Express $\cos 2\theta$ in terms of $\cos\theta$ and $\sin\theta$.

Solution

$$\begin{aligned}\cos 2\theta &= \operatorname{Re} e^{i2\theta} = \operatorname{Re}(\cos\theta + i\sin\theta)^2 \\ &= \operatorname{Re}(\cos^2\theta + 2i\cos\theta\sin\theta + i^2\sin^2\theta) = \cos^2\theta - \sin^2\theta.\end{aligned}$$

We can now give the definition of the exponential e^z for every complex number $z = x + iy$.

Definition 3.4. Let $z = x + iy$ be a complex number. Then we define e^z by

$$e^z = e^{x+iy} = e^x e^{iy} = e^x(\cos y + i\sin y).$$

Let $z = re^{i\theta}$ be a nonzero complex number. Then for every positive integer n, the nth power z^n of z is given by

$$z^n = r^n e^{in\theta} = r^n(\cos n\theta + i\sin n\theta).$$

Is there a formula for all the nth roots of z? To solve this problem, let $\zeta = \sqrt[n]{z}$. Suppose that $\zeta = \rho e^{i\phi}$. Then

$$\zeta^n = z \Rightarrow \rho^n e^{in\phi} = re^{i\theta}.$$

So,

$$\rho^n = r$$

and

$$\phi = \frac{1}{n}(\theta + 2k\pi),$$

where $k = 0, 1, 2, \ldots, n-1$. Thus, there are n nth roots of z given by

$$\zeta_k = \sqrt[n]{r}e^{i(\theta+2k\pi)/n}, \quad k = 0, 1, 2, \ldots, n-1.$$

Example 3.5. Find the four fourth roots of unity, *i.e.*, $\sqrt[4]{1}$.

Solution Since $1 = 1e^0$, we get $r = 1$ and $\theta = 0$. Thus, the four fourth roots of unity are

$$\zeta_k = e^{i(2k\pi)/4} = e^{ik\pi/2}, \quad k = 0, 1, 2, 3.$$

So,

$$\zeta_0 = 1,\ \zeta_1 = e^{i\pi/2} = i,\ \zeta_2 = e^{i\pi} = -1,\ \zeta_3 = e^{i3\pi/2} = -i.$$

Exercises

(1) Compute $\sum_{n=0}^{100} i^n$.
(2) Find all the fifth roots of unity. In other words, solve the equation $z^5 = 1$ for all complex numbers z.
(3) Solve the equation $(z+1)^5 = z^5$ for all complex numbers z.

Chapter 4

Functions of a Complex Variable

We need to study sets in the complex plane $\mathbb{C}$ because they come up as domains of functions.

Definition 4.1. Let $z_0 \in \mathbb{C}$ and let ρ be a positive number. Then we define the set $D(z_0, \rho)$ by

$$D(z_0, \rho) = \{z \in C : |z - z_0| < \rho\}.$$

We call $D(z_0, \rho)$ the open disk with center z_0 and radius ρ. We also call it a neighborhood of z_0.

Definition 4.2. Let G be a subset of $\mathbb{C}$. Suppose that for all $z_0 \in G$, there is an open disk centered at z_0 and contained in G. Then we say that G is an open set.

Example 4.3. Determine whether each of the following sets is an open set.

(1) $\mathbb{C}$
(2) $\{z \in \mathbb{C} : |z| < 1\}$
(3) $\{z \in \mathbb{C} : \operatorname{Im} z \geq 0\}$.

Solution The first and second sets are open. The third one is not.

Definition 4.4. Let $S \subseteq \mathbb{C}$. Suppose that every pair of points z_1 and z_2 in S can be joined by a polygonal line that lies entirely in S. Then we say that S is connected.

Example 4.5. Determine whether each of the following sets is connected.

(1) $\{z \in \mathbb{C} : 1 < |z| < 2\}$
(2) $\{z \in \mathbb{C} : \operatorname{Re} z < 1\} \cup \{z \in \mathbb{C} : \operatorname{Re} z > 2\}$
(3) $\{z \in \mathbb{C} : 0 < |z| \leq 1\}$.

Solution The first and the third are connected. The second one is not.

Here come the sets that are most important for us.

Definition 4.6. A subset D of $\mathbb{C}$ is said to be a domain if it is open and connected.

Example 4.7. Determine whether or not each set in Examples 4.3 and 4.5 is a domain.

Solution $\mathbb{C}$, $\{z \in \mathbb{C} : |z| < 1\}$ and $\{z \in \mathbb{C} : 1 < |z| < 2\}$ are domains. The others are not.

Let $S \subset \mathbb{C}$. Let $z_0 \in \mathbb{C}$ be such that every neighborhood of z_0 intersects S and $\mathbb{C} - S$. Then we call z_0 a boundary point of S.

Example 4.8. Find all the boundary points of each of the following sets.

(1) $\mathbb{C} - \{0\}$
(2) $\{z \in \mathbb{C} : 1 < |z| \leq 2\}$.

Solution For the first part, 0 is a boundary point. For the second part, the boundary points are given by $\{z \in \mathbb{C} : |z| = 1\} \cup \{ z \in \mathbb{C} : |z| = 2\}$.

Definition 4.9. Let S be a domain with some, none or all of its boundary points. Then we call S a region.

Let $S \subseteq \mathbb{C}$ be a region and let f be a function mapping S into $\mathbb{C}$. We find it occasionally convenient to write

$$w = f(z), \quad z \in S,$$

for the function f. If we write $z = x + iy$ and $w = u + iv$ in terms of the real and imaginary parts, then

$$w = u(x, y) + iv(x, y), \quad z = x + iy \in S,$$

and $u(x, y)$ and $v(x, y)$ are two real-valued functions of two real variables x and y on S.

Example 4.10. Express the function $w = f(z) = z^2 + zi$, $z \in \mathbb{C}$, in terms of its real and imaginary parts.

Solution Let $z = x + iy$. Then for all $z = x + iy \in \mathbb{C}$,

$$\begin{aligned} w = f(z) &= z^2 + zi = (x + iy)^2 + i(x + iy) \\ &= x^2 + 2ixy - y^2 + ix - y = (x^2 - y^2 - y) + i(2xy + x). \end{aligned}$$

We end this section with two concepts, which students should have come across in their calculus courses.

Definition 4.11. Let f be a function defined on some neighborhood of z_0 except possibly at z_0. Let $w_0 \in \mathbb{C}$. Suppose that for every positive number ε, there exists a positive number δ such that

$$0 < |z - z_0| < \delta \Rightarrow |f(z) - w_0| < \varepsilon.$$

Then we say that

$$\lim_{z \to z_0} f(z) = w_0$$

or

$$f(z) \to w_0$$

as $z \to z_0$.

Definition 4.12. Let f be a function defined on a neighborhood of z_0. Then f is continuous at z_0 if

$$\lim_{z \to z_0} f(z) = f(z_0).$$

Remark 4.13. Limits and continuity of functions of a complex variable are exactly the same as those encountered in first and second year calculus courses and hence we stop short of the repetitions.

Exercises

(1) Draw the set described by each of the following relations.

- (a) $\operatorname{Im} z > 1$
- (b) $\operatorname{Re} z \neq 0$
- (c) $|z - 2i| \leq 2$
- (d) $1 < |z| < 2$
- (e) $\operatorname{Im} z < |z|$
- (f) $|z - 1| < |z + i|$
- (g) $|z^2 - 1| < 1$.

(2) Which of the sets in the preceding exercise are domains?

Chapter 5

Holomorphic Functions and the Cauchy–Riemann Equations

In this book we study complex-valued functions defined on subsets of the complex plane $\mathbb{C}$. The subsets that are of most interest to us are known as domains. Domains are open and connected subsets of $\mathbb{C}$. A domain with some or all of its boundary points is known as a region.

Complex analysis can be characterized as the study of holomorphic functions. To see what holomorphic functions are, let us begin with some very familiar notions.

Definition 5.1. Let $w = f(z)$ be a complex-valued function defined on a neighborhood of z_0. Suppose that $\lim_{\Delta z \to 0} \frac{f(z_0+\Delta z)-f(z_0)}{\Delta z}$ exists. Then we say that f is differentiable at z_0. We call the limit the derivative of f at z_0 and we denote it by $f'(z_0)$.

Remark 5.2. It is important to realize that Δz is the notation for a number. It is not the product of Δ and z.

Example 5.3. Let $w = f(z) = z^2$, $z \in \mathbb{C}$. Find $f'(z)$.

Solution Let $z \in \mathbb{C}$. Then

$$\begin{aligned} f'(z) &= \lim_{\Delta z \to 0} \frac{f(z+\Delta z) - f(z)}{\Delta z} \\ &= \lim_{\Delta z \to 0} \frac{(z+\Delta z)^2 - z^2}{\Delta z} \\ &= \lim_{\Delta z \to 0} \frac{z^2 + 2z\,\Delta z + (\Delta z)^2 - z^2}{\Delta z} \\ &= \lim_{\Delta z \to 0} \frac{2z\,\Delta z + (\Delta z)^2}{\Delta z} \\ &= \lim_{\Delta z \to 0} (2z + \Delta z) = 2z. \end{aligned}$$

The rules for differentiation and for finding derivatives are the same as those in calculus courses.

Example 5.4. Prove that $w = f(z) = \overline{z}$, $z \in \mathbb{C}$, where $\overline{z}$ is the complex conjugate of z, is nowhere differentiable.

Solution Let $z \in \mathbb{C}$. Then

$$f'(z) = \lim_{\Delta z \to 0} \frac{f(z + \Delta z) - f(z)}{\Delta z} = \lim_{\Delta z \to 0} \frac{\overline{z + \Delta z} - \overline{z}}{\Delta z} = \lim_{\Delta z \to 0} \frac{\overline{\Delta z}}{\Delta z}.$$

Let $\Delta z \to 0$ along the x-axis. Then $\Delta z = \Delta x$ and

$$f'(z) = \lim_{\Delta x \to 0} \frac{\Delta x}{\Delta x} = 1.$$

Let $\Delta z \to 0$ along the y-axis. Then $\Delta z = i\Delta y$ and

$$f'(z) = \lim_{\Delta y \to 0} \frac{-i\Delta y}{i\Delta y} = -1.$$

Thus, $f'(z)$ does not exist.

Here comes the definition of a holomorphic function.

Definition 5.5. Let $w = f(z)$ be a complex-valued function defined on an open subset G of $\mathbb{C}$. Suppose that f is differentiable at every point in G. Then we say that f is holomorphic on G.

Definition 5.6. A holomorphic function on $\mathbb{C}$ is said to be an entire function.

Example 5.7. $w = f(z) = z^2$ is entire.

Remark 5.8. It is important to emphasize the fact that holomorphicity is a property defined on an open set, while differentiability may hold at one point only. However, we still say that f is holomorphic at the point z_0 when f is holomorphic on some neighborhood of z_0.

The real and imaginary parts of a holomorphic function are intimately related by the so-called Cauchy–Riemann equations, which we now derive.

Let $w = f(z) = u(x, y) + iv(x, y)$ be differentiable at the point $z_0 = x_0 + iy_0$. Let $\Delta z \to 0$ along the x-axis. Then $\Delta z = \Delta x$ and

$$
\begin{aligned}
&f'(z_0)\\
&= \lim_{\Delta z \to 0} \frac{f(z_0 + \Delta z) - f(z_0)}{\Delta z}\\
&= \lim_{\Delta x \to 0} \frac{u(x_0 + \Delta x, y_0) + iv(x_0 + \Delta x, y_0) - u(x_0, y_0) - iv(x_0, y_0)}{\Delta x}\\
&= \lim_{\Delta x \to 0} \frac{u(x_0 + \Delta x, y_0) - u(x_0, y_0)}{\Delta x} + i \lim_{\Delta x \to 0} \frac{v(x_0 + \Delta x, y_0) - v(x_0, y_0)}{\Delta x}\\
&= \frac{\partial u}{\partial x}(x_0, y_0) + i\frac{\partial v}{\partial x}(x_0, y_0).
\end{aligned}
$$

Now, let $\Delta z \to 0$ along the y-axis. Then $\Delta z = i\Delta y$ and

$$
\begin{aligned}
&f'(z_0)\\
&= \lim_{\Delta y \to 0} \frac{u(x_0, y_0 + \Delta y) + iv(x_0, y_0 + \Delta y) - u(x_0, y_0) - iv(x_0, y_0)}{i\Delta y}\\
&= \lim_{\Delta y \to 0} \frac{u(x_0, y_0 + \Delta y) - u(x_0, y_0)}{i\Delta y} + i \lim_{\Delta y \to 0} \frac{v(x_0, y_0 + \Delta y) - v(x_0, y_0)}{i\Delta y}\\
&= -i\frac{\partial u}{\partial y}(x_0, y_0) + \frac{\partial v}{\partial y}(x_0, y_0).
\end{aligned}
$$

Thus, we get

$$
\begin{cases}
\frac{\partial u}{\partial x}(x_0, y_0) = \frac{\partial v}{\partial y}(x_0, y_0),\\
\frac{\partial u}{\partial y}(x_0, y_0) = -\frac{\partial v}{\partial x}(x_0, y_0).
\end{cases}
$$

So, at $z_0 = x_0 + iy_0$, u and v satisfy a first order system of partial differential equations known as the Cauchy–Riemann equations.

Remark 5.9. If $w = f(z) = u(x, y) + iv(x, y)$ is differentiable at the point $z_0 = x_0 + iy_0$, then

$$f'(z_0) = \frac{\partial u}{\partial x}(x_0, y_0) + i\frac{\partial v}{\partial x}(x_0, y_0) = \frac{\partial v}{\partial y}(x_0, y_0) - i\frac{\partial u}{\partial y}(x_0, y_0).$$

Theorem 5.10. *If $w = f(z) = u(x, y) + iv(x, y)$ is differentiable at z_0, then the functions u and v satisfy the Cauchy–Riemann equations at z_0. Moreover, if f is holomorphic on an open set G, then the functions u and v satisfy the Cauchy–Riemann equations at every point in G.*

The following example, which we leave as an exercise, tells us that the converse to Theorem 5.10 is false.

Example 5.11. Let

$$w = f(z) = \begin{cases} \frac{x^{4/3}y^{5/3}+ix^{5/3}y^{4/3}}{x^2+y^2}, & z \neq 0, \\ 0, & z = 0. \end{cases}$$

Then the real and imaginary parts of f satisfy the Cauchy–Riemann equations at $z = 0$, but f is not differentiable at $z = 0$.

The converse is almost true. To state a partial converse, we need a definition. A complex-valued function

$$w = f(z) = u(x, y) + iv(x, y)$$

on an open set G is said to be C^1 on G if u, v and all their partial derivatives of first order are continuous on G.

Theorem 5.12. *Let $w = f(z) = u(x, y) + iv(x, y)$ be a function defined on an open set G containing the point z_0. If f is C^1 on a neighborhood of z_0 and the functions u and v satisfy the Cauchy–Riemann equations at z_0, then f is differentiable at z_0. Moreover, if f is C^1 on G and the functions u and v satisfy the Cauchy–Riemann equations at all points in G, then f is holomorphic on G.*

Proof It is enough to prove the first part. To that end, we only need to prove that $\lim_{h\to 0} \frac{f(z_0+h)-f(z_0)}{h}$ exists. Write $z_0 = x_0+iy_0$ and $h = h_1+ih_2$. Then for all nonzero complex numbers h with $z_0 + h \in G$,

$$\begin{aligned} &\frac{f(z_0 + h) - f(z_0)}{h} \\ &= \frac{u(x_0 + h_1, y_0 + h_2) - u(x_0, y_0)}{h} + i\frac{v(x_0 + h_1, y_0 + h_2) - v(x_0, y_0)}{h}. \end{aligned}$$

Since f is C^1 on a neighborhood of z_0, we get

$$\begin{aligned} &u(x_0 + h_1, y_0 + h_2) - u(x_0, y_0) \\ &= \frac{\partial u}{\partial x}(x_0, y_0)h_1 + \frac{\partial u}{\partial y}(x_0, y_0)h_2 + |h|\psi_1(h) \end{aligned}$$

and

$$\begin{aligned} &v(x_0 + h_1, y_0 + h_2) - v(x_0, y_0) \\ &= \frac{\partial v}{\partial x}(x_0, y_0)h_1 + \frac{\partial v}{\partial y}(x_0, y_0)h_2 + |h|\psi_2(h), \end{aligned}$$

where $\psi_1(h) \to 0$ and $\psi_2(h) \to 0$ as $|h| \to 0$. Then we can apply the Cauchy–Riemann equations and we get

$$\begin{aligned}\frac{f(z_0+h)-f(z_0)}{h} &= \frac{\partial u}{\partial x}(x_0,y_0) + i\frac{\partial v}{\partial x}(x_0,y_0) + \frac{|h|(\psi_1(h)+i\psi_2(h))}{h} \\ &\to \frac{\partial u}{\partial x}(x_0,y_0) + i\frac{\partial v}{\partial x}(x_0,y_0)\end{aligned}$$

as $h \to 0$.

□

Example 5.13. Prove that $w = f(z) = \overline{z}$, $z \in \mathbb{C}$, is nowhere differentiable.

Solution $w = f(z) = x - iy$, $z = x + iy \in \mathbb{C}$. Then $u(x,y) = x$ and $v(x,y) = -y$. Now,

$$\frac{\partial u}{\partial x} = 1, \quad \frac{\partial v}{\partial y} = -1,$$

and

$$\frac{\partial u}{\partial y} = 0, \quad \frac{\partial v}{\partial x} = 0.$$

Therefore the Cauchy–Riemann equations are not valid at any point in $\mathbb{C}$. Thus, $w = f(z) = \overline{z}$, $z \in \mathbb{C}$, is nowhere differentiable.

Example 5.14. Prove that $w = f(z) = x^2 + y + i(y^2 - x)$, $z = x + iy \in \mathbb{C}$, is nowhere holomorphic.

Solution Let $u(x,y) = x^2 + y$ and $v(x,y) = y^2 - x$. Then

$$\frac{\partial u}{\partial x} = 2x, \quad \frac{\partial v}{\partial y} = 2y,$$

and

$$\frac{\partial u}{\partial y} = 1, \quad \frac{\partial v}{\partial x} = -1.$$

Thus, the Cauchy–Riemann equations are only valid on the line $y = x$. Therefore $w = f(z)$ is differentiable only at the points in the line $y = x$. Hence $w = f(z)$ is nowhere holomorphic.

Example 5.15. Prove that the function

$$w = f(z) = e^x \cos y + ie^x \sin y, \quad z = x + iy \in \mathbb{C},$$

is entire and find $f'(z)$ for every $z \in \mathbb{C}$.

Solution For all $z = x + iy \in \mathbb{C}$, let

$$u(x, y) = e^x \cos y$$

and

$$v(x, y) = e^x \sin y.$$

Then

$$\frac{\partial u}{\partial x} = e^x \cos y, \quad \frac{\partial v}{\partial y} = e^x \cos y,$$

and

$$\frac{\partial u}{\partial y} = -e^x \sin y, \quad \frac{\partial v}{\partial x} = e^x \sin y.$$

The first partial derivatives are continuous and the Cauchy–Riemann equations are fulfilled at every point $z = x + iy$ in $\mathbb{C}$. Therefore $w = f(z)$ is entire and

$$f'(z) = \frac{\partial u}{\partial x} + i\frac{\partial v}{\partial x} = e^x \cos y + ie^x \sin y = f(z)$$

for all $z \in \mathbb{C}$.

Exercises

(1) For the function

$$w = f(z) = x^3 - y^3 - 3ixy, \quad z = x + iy \in \mathbb{C},$$

find the set of all complex numbers z on which f is differentiable and determine the set on which the function f is holomorphic.

(2) Prove that the real and imaginary parts of the function in Example 5.11 satisfy the Cauchy–Riemann equations at $z = 0$, but the function is not differentiable at $z = 0$.

(3) Let $\overline{\partial}$ be the partial differential operator on $\mathbb{R}^2$ given by

$$\overline{\partial} = \frac{\partial}{\partial x} + i\frac{\partial}{\partial y}.$$

Let $w = f(z)$ be a C^1 complex-valued function on an open set G. Prove that f is holomorphic on G if and only if

$$(\overline{\partial} f)(z) = 0, \quad z \in G.$$

(4) Let G be an open subset of the complex plane $\mathbb{C}$ such that

$$z \in G \Rightarrow \overline{z} \in G.$$

Then we call G a symmetric open set. Prove that if f is a holomorphic function on a symmetric open set G, then the function $w = \overline{f(\overline{z})}$ is also a holomorphic function on G.

(5) Let $w = f(z)$ be a holomorphic function on a domain D such that $f'(z) = 0$ for all z in D. Prove that f is a constant function on D.

(6) Let $w = f(z)$ be a holomorphic function on a domain D. Prove that if $f(z)$ is real-valued for all $z \in D$, then f is a constant function on D.

(7) Let $w = f(z)$ be a function on a domain D such that f and the complex conjugate $\overline{f}$ of f are holomorphic on D. Prove that f is a constant function on D.

(8) Let $w = f(z)$ be a function on a domain D such that f and the absolute value $|f|$ of f are holomorphic on D. Prove that f is a constant function on D.

Chapter 6

The Exponential, Trigonometric and Hyperbolic Functions

We begin with the definition of the function e^z for all $z \in \mathbb{C}$. Guided by the properties of the exponential function e^x for all $x \in \mathbb{R}$, we expect the function e^z to be an entire function such that

$$e^{z_1}e^{z_2} = e^{z_1+z_2}, \quad z_1, z_2 \in \mathbb{C},$$

and

$$e^0 = 1.$$

To do this, let $z = x + iy \in \mathbb{C}$. Then

$$e^z = e^{x+iy} = e^x e^{iy} = e^x(c(y) + is(y)) = e^x c(y) + ie^x s(y),$$

where $c(y)$ and $s(y)$ are to be determined. First, we note that

$$1 = e^{i0} = c(0) + is(0).$$

Hence $c(0) = 1$ and $s(0) = 0$. Let

$$u(x, y) = e^x c(y)$$

and

$$v(x, y) = e^x s(y)$$

for all $z = x + iy \in \mathbb{C}$. Now,

$$\frac{\partial u}{\partial x} = e^x c(y), \quad \frac{\partial v}{\partial y} = e^x s'(y), \quad z = x + iy \in \mathbb{C},$$

and

$$\frac{\partial u}{\partial y} = e^x c'(y), \quad \frac{\partial v}{\partial x} = e^x s(y), \quad z = x + iy \in \mathbb{C}.$$

By the Cauchy–Riemann equations, we get

$$e^x c(y) = e^x s'(y), \quad x, y \in \mathbb{R},$$

and

$$e^x c'(y) = -e^x s(y), \quad x, y \in \mathbb{R}.$$

So, we get the system of ordinary differential equations

$$\begin{cases} s' = c, \\ c' = -s. \end{cases}$$

Thus, we get

$$\begin{cases} s'' + s = 0, \\ s(0) = 0. \end{cases}$$

Therefore $s(y) = \sin y$ and $c(y) = \cos y$ for all $y \in \mathbb{R}$. This gives Euler's identity to the effect that

$$e^{iy} = \cos y + i \sin y, \quad y \in \mathbb{R},$$

and the very important formula

$$e^z = e^{x+iy} = e^x(\cos y + i \sin y), \quad z = x + iy \in \mathbb{C},$$

which is the same as the function studied in Example 5.15. Therefore

$$\frac{d}{dz}(e^z) = e^z, \quad z \in \mathbb{C}.$$

Example 6.1. Find all zeros of $e^z = 1$.

Solution Let $z = x + iy$. Then

$$\begin{cases} e^x \cos y = 1, \\ e^x \sin y = 0. \end{cases}$$

Thus, from the second equation,

$$y = k\pi, \quad k = 0, \pm 1, \pm 2, \ldots.$$

If k is odd, then putting $y = k\pi$ in the first equation, we get

$$e^x \cos k\pi = 1$$

or

$$-e^x = 1,$$

which is impossible. So, we only need to look at

$$y = 2k\pi, \quad k = 0, \pm 1, \pm 2, \ldots.$$

Putting these values in the first equation, we get $e^x = 1$. So, $x = 0$. Therefore

$$z = iy = 2k\pi i, \quad k = 0, \pm 1, \pm 2, \ldots.$$

Example 6.2. Prove that the exponential function $w = e^z$, $z \in \mathbb{C}$, is periodic with period $2\pi i$, *i.e.*, $e^{z+2\pi i} = e^z$, $z \in \mathbb{C}$.

Solution Since

$$e^{z+2\pi i}e^{-z} = e^{2\pi i} = 1, \quad z \in \mathbb{C},$$

we get

$$e^{z+2\pi i} = e^z, \quad z \in \mathbb{C},$$

as required.

By Euler's identity, we get

$$e^{\pm iy} = \cos y \pm i \sin y, \quad y \in \mathbb{R}.$$

Therefore

$$\cos y = \frac{e^{iy} + e^{-iy}}{2}$$

and

$$\sin y = \frac{e^{iy} - e^{-iy}}{2i}$$

for all y in $\mathbb{R}$. These discussions lead us to define trigonometric functions of a complex variable.

Definition 6.3. For every complex number z, we define $\cos z$ and $\sin z$ by

$$\cos z = \frac{e^{iz} + e^{-iz}}{2}$$

and

$$\sin z = \frac{e^{iz} - e^{-iz}}{2i}.$$

Remark 6.4. For every complex number z, we define $\tan z$, $\sec z$, $\csc z$ and $\cot z$ as in calculus. The derivatives and the trigonometric identities for these six functions are also the same as in calculus.

We end this chapter with a brief discussion of hyperbolic functions.

Definition 6.5. For every complex number z, we define $\cosh z$ and $\sinh z$ by

$$\cosh z = \frac{e^z + e^{-z}}{2}$$

and

$$\sinh z = \frac{e^z - e^{-z}}{2}.$$

As in the case of trigonometric functions, we can define $\tanh z$, $\operatorname{sech} z$ $\operatorname{csch} z$ and $\coth z$ for every complex number z. Their derivatives and the familiar identities can be derived in exactly the same way as in calculus.

Exercises

(1) Find the real part, the imaginary part and the absolute value of e^{e^z}, where z is a complex number.
(2) Are the functions $\sin z$ and $\cos z$ periodic functions on $\mathbb{C}$?
(3) Is it true that $|\sin z| \leq 1$ for all $z \in \mathbb{C}$? Explain your answer. (Hint: Try $z = iy$, $y \in \mathbb{R}$.)

Chapter 7

Logarithms, Complex Powers, Branches and Cuts

A very interesting function in complex analysis is the logarithmic function

$$w = \log z,$$

which we now introduce.

As in calculus, we want to define $\log z$ as the inverse of the exponential function studied in the preceding chapter. Thus,

$$w = \log z \Leftrightarrow z = e^w.$$

As usual, we write $w = u + iv$ and $z = x + iy$. Since $|e^w| = e^u \neq 0$, we see that $z \neq 0$. So, we can write z in polar form as

$$z = re^{i\theta}.$$

Therefore

$$re^{i\theta} = e^u e^{iv}.$$

We get $r = e^u$ or $u = \ln r$. We also get

$$v = \theta = \arg z.$$

So, we have the formula

$$\log z = \ln |z| + i \arg z, \quad z \neq 0.$$

Remark 7.1. For every nonzero complex number z,

$$\log z = \ln |z| + i\,(\theta + 2k\pi), \quad k = 0, \pm 1, \pm 2, \ldots,$$

where θ is any value or branch of $\arg z$.

Definition 7.2. Let $\tau \in \mathbb{R}$. Then for every nonzero complex number z, we define the branch $\log_\tau z$ of $\log z$ corresponding to τ by

$$\log_\tau z = \ln |z| + i \arg_\tau z.$$

If we let $\tau = -\pi$, then we write $\operatorname{Log} z$ for $\log_{-\pi} z$ and we call $\operatorname{Log} z$ the principal logarithm of z.

Thus, it is important to note that

$$\operatorname{Log} z = \ln|z| + i \arg_{-\pi} z = \ln|z| + i \operatorname{Arg} z, \quad z \neq 0.$$

Example 7.3. Compute $\log(1+i)$, $\operatorname{Log}(1+i)$ and $\log_\pi(1+i)$.

Solution Let $z = 1+i$. Then $|z| = \sqrt{2}$ and

$$\arg z = \frac{\pi}{4} + 2k\pi, \quad k = 0, \pm 1, \pm 2, \ldots.$$

Thus,

$$\log(1+i) = \ln\sqrt{2} + i\left(\frac{\pi}{4} + 2k\pi\right), \quad k = 0, \pm 1, \pm 2, \ldots,$$

$$\operatorname{Log}(1+i) = \ln\sqrt{2} + i\frac{\pi}{4}$$

and

$$\log_\pi(1+i) = \ln\sqrt{2} + i\frac{9\pi}{4}.$$

Theorem 7.4. *Let $\tau \in \mathbb{R}$. Then $w = f(z) = \log_\tau z$ is holomorphic on the domain $\mathbb{C}_\tau^\circ$ given by*

$$\mathbb{C}_\tau^\circ = \{(r,\theta) : r > 0, \tau < \theta < \tau + 2\pi\},$$

and

$$\frac{d}{dz}(\log_\tau z) = \frac{1}{z}, \quad z \in \mathbb{C}_\tau^\circ.$$

Proof Let $z_0 \in \mathbb{C}_\tau^\circ$. Then we need to prove that f is differentiable at z_0. Let $w_0 = \log_\tau z_0$. Then we need to prove that

$$\lim_{z \to z_0} \frac{f(z) - f(z_0)}{z - z_0} = \lim_{z \to z_o} \frac{w - w_0}{z - z_0} = \frac{1}{z_0}.$$

But

$$w = \log_\tau z \Leftrightarrow z = e^w$$

and

$$e^{w_0} = \lim_{w \to w_0} \frac{e^w - e^{w_0}}{w - w_0} = \lim_{w \to w_0} \frac{z - z_0}{w - w_0}.$$

Note that

$$\log_\tau z = \ln|z| + i \arg_\tau z \to \ln|z_0| + i \arg_\tau z_0 = \log_\tau z_0$$

as $z \to z_0$. Therefore $z \to z_0 \Rightarrow w \to w_0$. Moreover,

$$z \neq z_0 \Rightarrow w \neq w_0$$

because

$$w = w_0 \Rightarrow z = e^w = e^{w_0} = z_0.$$

Therefore

$$\lim_{z \to z_0} \frac{w - w_0}{z - z_0} = \lim_{w \to w_0} \frac{1}{\frac{z - z_0}{w - w_0}} = \frac{1}{e^{w_0}} = \frac{1}{z_0},$$

as required. □

Example 7.5. Find the domain on which the function

$$w = f(z) = \mathrm{Log}\,(3z - i)$$

is holomorphic. Compute $f'(z)$.

Solution Since f is the composition of Log and $3z - i$, f is holomorphic at all points z unless $3z - i \in (-\infty, 0]$. But it is easy to see that

$$\begin{aligned} 3z - i \in (-\infty, 0] &\Leftrightarrow 3z - i = x,\ x \leq 0, \\ &\Leftrightarrow 3z = x + i,\ x \leq 0, \\ &\Leftrightarrow z = \frac{x}{3} + \frac{i}{3},\ x \leq 0. \end{aligned}$$

Thus, f is holomorphic on the domain $\mathbb{C} - \left\{x + \frac{i}{3} : x \leq 0\right\}$. For a picture of the domain, see Figure 7.1.

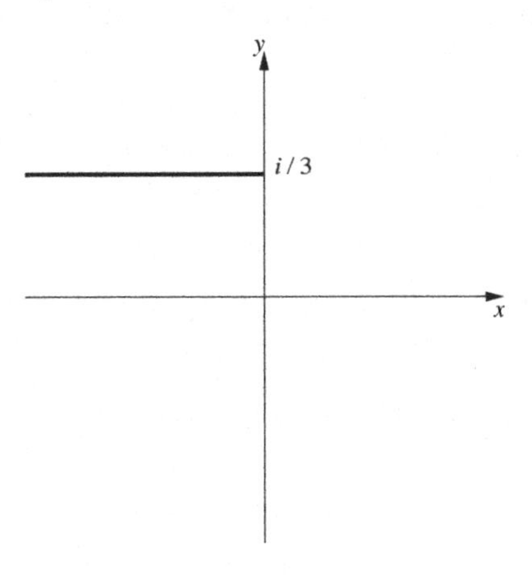

Fig. 7.1 Domain of f

Example 7.6. Find a branch of $w = f(z) = \log\,(z^3 - 2)$ that is holomorphic at $z = 0$. Find $f(0)$ and $f'(0)$.

Solution Let $\tau \in \mathbb{R}$ and take the branch $\log_\tau$. Then f is the composition of $\log_\tau$ and the function $z^3 - 2$. So, f is holomorphic at $z = 0$ only if -2 is not on the branch cut $\{(r, \tau) : r > 0\}$. So, we can use any branch

$$f(z) = \log_\tau(z^3 - 2), \quad z \in \mathbb{C}^\circ_\tau,$$

where the branch cut $\{(r, \tau) : r > 0\}$ is not equal to $(-\infty, 0)$. Also,

$$f(0) = \log_\tau(-2) = \ln|-2| + i\arg_\tau(-2) = \ln|2| + i\arg_\tau(-2).$$

and

$$f'(0) = \left.\frac{3z^2}{z^3 - 2}\right|_{z=0} = 0.$$

We can now come to a study of complex powers.

Definition 7.7. Let z be a nonzero complex number. Then for every complex number α, we define z^α by

$$z^\alpha = e^{\alpha \log z}.$$

Again, the multi-valuedness of z^α is an interesting issue.

Example 7.8. Find all the values of $(-2)^i$.

Solution We get

$$(-2)^i = e^{i\log(-2)} = e^{i(\ln|-2| + i\arg(-2))} = e^{i(\ln|2| + i(\pi + 2k\pi))} = e^{i\ln|2|}e^{-(2k+1)\pi}$$

for $k = 0, \pm 1, \pm 2, \ldots$.

Example 7.9. Find a branch of $w = f(z) = (z^2 - 1)^{1/2}$ that is holomorphic on $\{z \in \mathbb{C} : |z| > 1\}$.

Remark 7.10. Let us see whether or not we can use

$$f(z) = e^{\frac{1}{2}\mathrm{Log}(z^2 - 1)}.$$

Note that f is not holomorphic at all z with $z^2 - 1 \in (-\infty, 0]$. But

$$z^2 - 1 \in (-\infty, 0] \Leftrightarrow z^2 \le 1.$$

It is then easy to see that f is not holomorphic at every point in the "cross" given by the union of the y-axis and the interval $[-1, 1]$. See Figure 7.2. So, the branch $f(z) = e^{\frac{1}{2}\mathrm{Log}(z^2 - 1)}$ does not work.

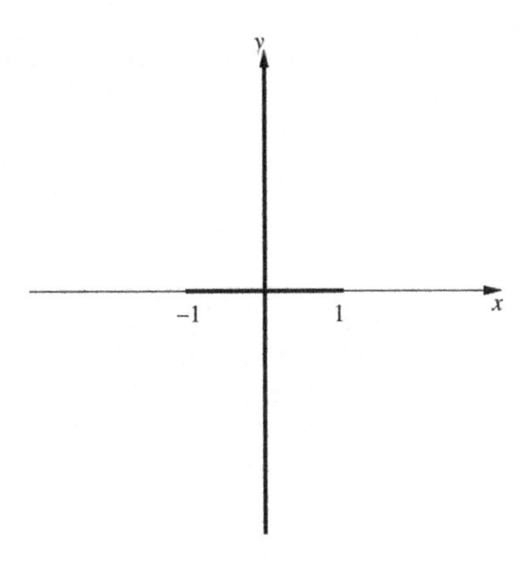

Fig. 7.2 The Cross

Solution of Example 7.9 The idea is to find a function $w = f(z)$ such that $w^2 = z^2 - 1$ and f is holomorphic on $\{z \in \mathbb{C} : |z| > 1\}$. But

$$w^2 = z^2 - 1 = z^2 \left(1 - \frac{1}{z^2}\right).$$

So,

$$w = z \left(1 - \frac{1}{z^2}\right)^{\frac{1}{2}}.$$

Now, we choose the branch

$$w = f(z) = ze^{\frac{1}{2}\operatorname{Log}\left(1 - \frac{1}{z^2}\right)}.$$

Then f is holomorphic at all z unless

$$1 - \frac{1}{z^2} \in (-\infty, 0],$$

which is the same as

$$\frac{1}{z^2} \geq 1.$$

Let $z \neq 0$. If we write $z = re^{i\theta}$, then

$$\frac{1}{z^2} = \frac{1}{r^2} e^{-2i\theta} \geq 1.$$

So, it is easy to see that

$$2\theta = 2k\pi, \quad k = 0, \pm 1, \pm 2, \ldots.$$

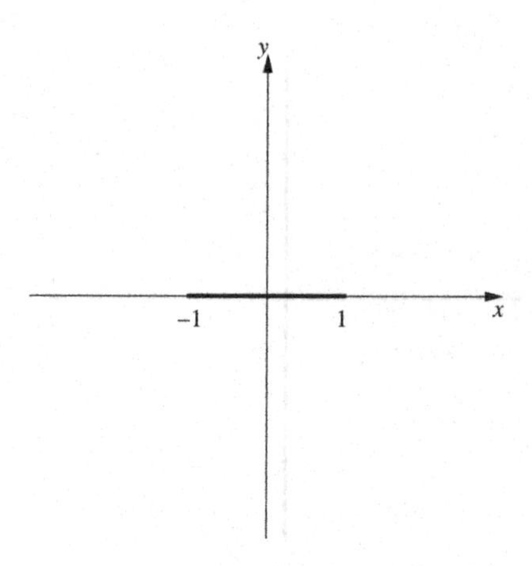

Fig. 7.3 The Forbidden Interval

Thus,

$$\theta = k\pi, \quad k = 0, \pm 1, \pm 2, \ldots,$$

which means that

$$z = re^{i\theta} = re^{ik\pi} = r(-1)^k, \quad k = 0, \pm 1, \pm 2, \ldots.$$

Therefore z is a real number in $[-1, 1]$. Hence we can conclude that f is holomorphic at all points not in the forbidden interval $[-1, 1]$. See Figure 7.3.

Exercises

(1) Find a branch of $\log(z^2 + 1)$ that is holomorphic at $z = -1$. What is the derivative of the branch at $z = -1$?

(2) Prove that there exists no function $F(z)$ holomorphic on the annulus $D = \{z \in \mathbb{C} : 1 < |z| < 2\}$ such that

$$F'(z) = \frac{1}{z}, \quad z \in D.$$

(3) Find the derivative of the principal branch of z^{1+i} at $z = i$.

(4) Find a branch of each of the following multi-valued functions that is holomorphic on the given domain.

   (a) $(z^2 - 1)^{1/2}$ on $\{z \in \mathbb{C} : |z| < 1\}$
   (b) $(z^2 - 1)^{1/2}$ on $\{z \in \mathbb{C} : |z| > 1\}$
   (c) $(z^2 + 1)^{1/2}$ on $\mathbb{C}$ cut along the imaginary axis from $-i$ to i

(d) $(z^5 - 1)^{1/5}$ on $\{z \in \mathbb{C} : |z| > 1\}$.

(5) Let $\tau \in \mathbb{R}$. Prove that the function $w = f(z) = \arg_\tau z$ is nowhere holomorphic on $\mathbb{C}_\tau^\circ$.

Chapter 8

Contour Integrals and Path Independence

Let γ be a curve in $\mathbb{C}$. Suppose that it is parametrized by a continuous complex-valued function $z = z(t)$ for all t in $[a, b]$ such that

(1) $z(t)$ has a continuous derivative on $[a, b]$,
(2) $z'(t) \neq 0$ for all t in $[a, b]$,
(3) $z(t)$ is one-to-one on $[a, b]$.

Then we call γ a smooth arc. It is worth pointing out that in Conditions (1) and (2),

$$z'(a) = \lim_{t \to a+} \frac{z(t) - z(a)}{t - a}$$

and

$$z'(b) = \lim_{t \to b-} \frac{z(t) - z(b)}{t - b}.$$

Furthermore, to say that $z(t)$ has a continuous derivative at a and b, we mean that $z'(a)$ and $z'(b)$ exist,

$$\lim_{t \to a+} z'(t) = z'(a)$$

and

$$\lim_{t \to b-} z'(t) = z'(b).$$

Suppose that γ is a curve with a parametrization $z = z(t)$ on $[a, b]$ satisfying Conditions (1) and (2), and such that $z(t)$ is one-to-one on $[a, b)$, $z(b) = z(a)$ and $z'(b) = z'(a)$. Then we call γ a smooth closed curve. A smooth curve is either a smooth arc or a smooth closed curve. A smooth curve is always oriented in the direction of increasing time t in $[a, b]$.

Definition 8.1. A contour Γ is either a single point z_0 or a finite sequence of smooth curves $\gamma_1, \gamma_2, \ldots, \gamma_n$ such that the terminal point of γ_k coincides with the initial point of γ_{k+1} for $k = 1, 2, \ldots, n-1$. We write

$$\Gamma = \gamma_1 + \gamma_2 + \cdots + \gamma_n.$$

Definition 8.2. Let $w = f(z)$ be a continuous complex-valued function on a smooth curve γ with parametrization $z = z(t)$, $t \in [a, b]$. Then we define the contour integral $\int_\gamma f(z)\, dz$ by

$$\int_\gamma f(z)\, dz = \int_a^b f(z(t)) z'(t)\, dt.$$

Definition 8.3. Let $w = f(z)$ be a continuous complex-valued function on a contour $\Gamma = \gamma_1 + \gamma_2 + \cdots + \gamma_n$. Then we define the contour integral $\int_\Gamma f(z)\, dz$ by

$$\int_\Gamma f(z)\, dz = \int_{\gamma_1} f(z)\, dz + \int_{\gamma_2} f(z)\, dz + \cdots + \int_{\gamma_n} f(z)\, dz.$$

Example 8.4. Let $z_0 \in \mathbb{C}$ and let n be an integer. Compute $\int_C (z - z_0)^n dz$, where C is the circle $|z - z_0| = r$ traversed once in the counterclockwise direction.

Solution A parametrization of C is

$$z = z(t) = z_0 + re^{it}, \quad t \in [0, 2\pi].$$

Thus,

$$\int_C (z - z_0)^n dz = \int_0^{2\pi} r^n e^{int} ire^{it} dt = ir^{n+1} \int_0^{2\pi} e^{i(n+1)t} dt.$$

But

$$\int_0^{2\pi} e^{i(n+1)t} dt = \begin{cases} \left. \frac{e^{i(n+1)t}}{i(n+1)} \right|_0^{2\pi}, & n \neq -1, \\ 2\pi, & n = -1. \end{cases}$$

Therefore

$$\int_C (z - z_0)^n dz = \begin{cases} 0, & n \neq -1, \\ 2\pi i, & n = -1. \end{cases}$$

The following estimate for contour integrals will be useful to us later in the book. We give it now as an exercise in contour integration.

Theorem 8.5. *Let $w = f(z)$ be a continuous complex-valued function on a contour Γ such that there exists a positive constant M for which*

$$|f(z)| \leq M, \quad z \in \Gamma.$$

Then

$$\left| \int_{\Gamma} f(z)\, dz \right| \leq ML,$$

where L is the length of the contour Γ.

Proof We give the proof when Γ is a smooth curve γ. Let

$$z = z(t), \quad a \leq t \leq b,$$

be a parametrization of γ. Then

$$\left| \int_{\gamma} f(z)\, dz \right| \leq \int_a^b |f(z(t))|\, |z'(t)|\, dt \leq M \int_a^b |z'(t)|\, dt.$$

Let $z(t) = x(t) + i\, y(t)$, $a \leq t \leq b$. Then

$$|z'(t)| = \sqrt{x'(t)^2 + y'(t)^2}, \quad a \leq t \leq b.$$

Thus,

$$\left| \int_{\gamma} f(z)\, dz \right| \leq M \int_a^b \sqrt{x'(t)^2 + y'(t)^2}\, dt = ML.$$

□

Example 8.6. Find an upper bound for $\left| \int_{\Gamma} \frac{e^z}{z^2+1} dz \right|$, where Γ is the circle $|z| = 2$ traversed once in the counterclockwise direction.

Solution On the circle $\{z \in \mathbb{C} : |z| = 2\}$,

$$|e^z| = e^x \leq e^2$$

and

$$|z^2 + 1| \geq |z|^2 - 1 = 4 - 1 = 3.$$

So, by Theorem 8.5,

$$\left| \int_{\Gamma} \frac{e^z}{z^2+1} dz \right| \leq \frac{e^2}{3} 4\pi = \frac{4\pi e^2}{3}.$$

It is sometimes difficult or even impossible to compute contour integrals using parametrizations. Given a contour integral on a certain contour, it is sometimes possible to replace the given contour by one on which the integral is easily computed. This is the idea behind the concept of path independence.

Theorem 8.7. *Let $w = f(z)$ be a continuous complex-valued function on a domain D such that f has an antiderivative F on D, i.e., $F'(z) = f(z)$ for all z in D. Then for every contour Γ in D with initial point z_I and terminal point z_T,*

$$\int_\Gamma f(z)\,dz = F(z_T) - F(z_I).$$

Note that the value of the integral depends only on the endpoints and not on the actual path joining them.

Proof of Theorem 8.7 We again assume that Γ is a smooth curve γ with parametrization $z = z(t)$, $a \le t \le b$. Then

$$\int_\gamma f(z)\,dz = \int_a^b f(z(t))z'(t)\,dt.$$

But, using the chain rule,

$$\frac{d}{dt}(F(z(t))) = F'(z(t))z'(t) = f(z(t))z'(t), \quad a \le t \le b.$$

Thus,

$$\begin{aligned}\int_\gamma f(z)\,dz &= \int_a^b \frac{d}{dt}(F(z(t)))\,dt = F(z(t))|_a^b \\ &= F(z(b)) - F(z(a)) = F(z_T) - F(z_I).\end{aligned}$$

Corollary 8.8. *Let $w = f(z)$ be a continuous complex-valued function on a domain D such that f has an antiderivative on D. Then*

$$\int_\Gamma f(z)\,dz = 0$$

for every closed contour Γ in D.

Example 8.9. Let Γ be the contour consisting of the upper semicircle with radius 1 and center at the origin and a line segment joining 1 and $2+i$. It is oriented as shown in Figure 8.1. Compute $\int_\Gamma \cos z\,dz$.

Solution $\cos z$ has an antiderivative $\sin z$ on $\mathbb{C}$. So,

$$\int_\Gamma \cos z\,dz = \sin(2+i) - \sin(-1).$$

Example 8.10. Let $w = f(z) = \frac{1}{z}$ for all z in $\mathbb{C} - \{0\}$. Does it have an antiderivative on $\mathbb{C} - \{0\}$?

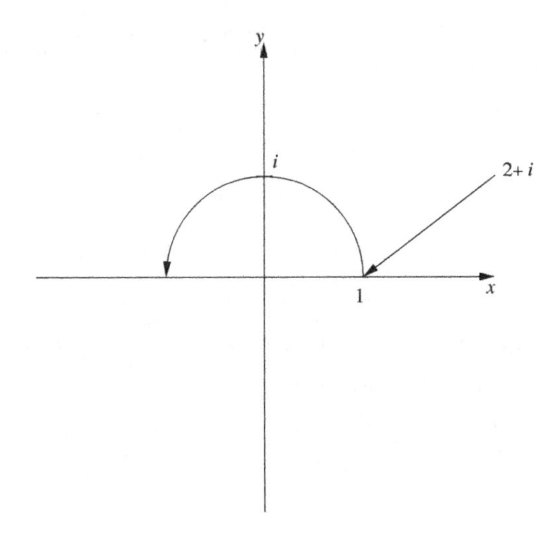

Fig. 8.1 The Contour Γ

Solution Let C be the unit circle with center at the origin traversed once in the counterclockwise direction. Its parametrization is $z = e^{it}$, $0 \leq t \leq 2\pi$. Then

$$\int_C \frac{1}{z}\,dz = \int_0^{2\pi} e^{-it} i\, e^{it}\,dt = 2\pi i.$$

Therefore, by Corollary 8.8, f has no antiderivative on $\mathbb{C} - \{0\}$.

The existence of an antiderivative and path independence in complex analysis mean the same thing. To make this precise, we have the following theorem.

Theorem 8.11. *Let $w = f(z)$ be a continuous complex-valued function on a domain D. Then the following statements are equivalent.*

(1) f has an antiderivative on D.
(2) $\int_\Gamma f(z)\,dz = 0$ for every closed contour Γ in D.
(3) $\int_{\Gamma_1} f(z)\,dz = \int_{\Gamma_2} f(z)\,dz$ for all contours Γ_1 and Γ_2 in D with the same initial and terminal points.

Proof By Corollary 8.8, Part (1) $\Rightarrow$ Part (2). To see that Part (2) $\Rightarrow$ Part (3), let Γ_1 and Γ_2 be contours in D with the same initial and terminal points as shown in Figure 8.2. Let Γ be the contour given by $\Gamma = \Gamma_1 - \Gamma_2$,

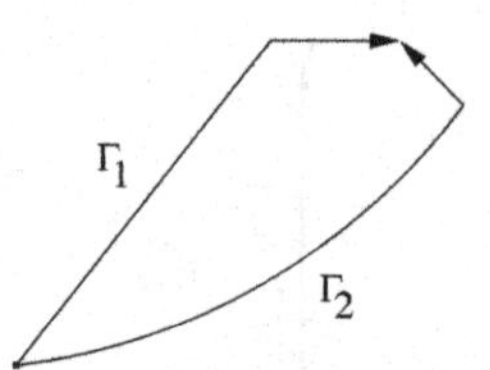

Fig. 8.2 Two Contours with Same Endpoints

where $-\Gamma_2$ is the same contour as Γ_2 except that the direction is reversed. Then Γ is a closed contour. Therefore

$$0 = \int_{\Gamma} f(z)\,dz = \int_{\Gamma_1} f(z)\,dz + \int_{-\Gamma_2} f(z)\,dz = \int_{\Gamma_1} f(z)\,dz - \int_{\Gamma_2} f(z)\,dz.$$

So,

$$\int_{\Gamma_1} f(z)\,dz = \int_{\Gamma_2} f(z)\,dz.$$

It remains to prove that Part (3) $\Rightarrow$ Part (1). Let $z_0 \in D$. Then for every point z in D, we join z_0 and z by a polygonal line Γ with initial point z_0 and terminal point z as in Figure 8.3. Let F be the function on D defined by

$$F(z) = \int_{z_0}^{z} f(w)\,dw, \quad z \in D.$$

Now,

$$F'(z) = \lim_{h\to 0} \frac{F(z+h) - F(z)}{h} = \lim_{h\to 0} \int_{z}^{z+h} f(w)\,dw.$$

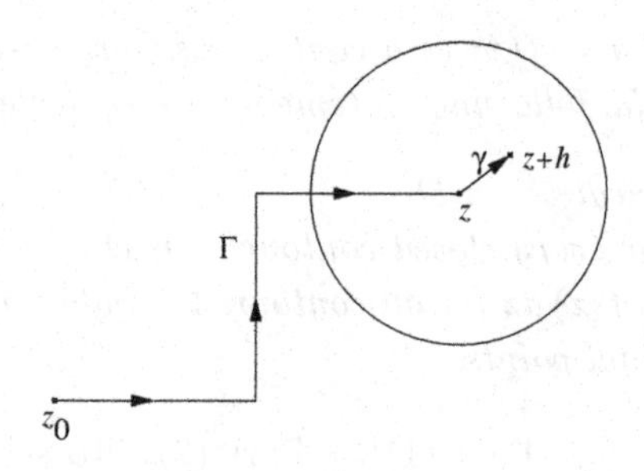

Fig. 8.3 Γ and γ

We let γ be the line segment joining z and $z+h$. Then it has the parametrization $w = z + th$, $0 \leq t \leq 1$. Thus,

$$F'(z) = \lim_{h\to 0} \frac{1}{h} \int_0^1 f(z+th)\, h\, dt = \lim_{h\to 0} \int_0^1 f(z+th)\, dt = f(z).$$

□

Exercises

(1) **(Wallis' Formula)** Let n be a nonnegative integer. Prove that

$$\frac{1}{2\pi} \int_0^{2\pi} (2\cos t)^{2n}\, dt = \frac{(2n)!}{(n!)^2}$$

by integrating the function $\frac{1}{z}\left(z + \frac{1}{z}\right)^{2n}$ around the unit circle centered at the origin and oriented once in the counterclockwise direction.

(2) Prove that

$$\left| \int_\Gamma \operatorname{Log} z\, dz \right| \leq \frac{\pi^2}{4},$$

where Γ is the quarter unit circle centered at the origin, lying in the first quadrant and oriented once in the counterclockwise direction.

(3) Prove that

$$\left| \int_\Gamma e^{\sin z} dz \right| \leq 1,$$

where Γ is the line segment directed from 0 to i.

(4) Let C be a circle in the complex plane such that the origin is neither on C nor inside C. Suppose that C is oriented once in the counterclockwise direction. Compute $\int_C \frac{1}{z} dz$.

Chapter 9

Cauchy's Integral Theorems

A criterion for a continuous complex-valued function $w = f(z)$ on a domain D to have an antiderivative on D is that the integral of f on every closed contour in D is equal to zero. How is it possible to check that the integral on every closed contour is equal to zero? It seems to be a tall order indeed. We show in this chapter that for holomorphic functions $w = f(z)$ on simply connected domains D, $\int_\Gamma f(z)\,dz = 0$ for every closed contour Γ in D. This surprising result is known as Cauchy's integral theorem. We give in the appendix of this chapter a proof of this result when the closed contour is a rectangle. The full-blown case is a consequence of a version of Cauchy's integral theorem based on continuous deformations of closed contours. We state this more general version of Cauchy's integral theorem without proof. Then we work on some examples to appreciate the power of Cauchy's integral theorem.

Continuous deformations of closed contours are easy to visualize. To wit, let D be a domain in the complex plane. Let Γ_0 and Γ_1 be closed contours in D. Suppose that Γ_0 can be continuously moved in the plane in such a manner that it coincides with Γ_1 in position and direction without getting out of D. Then we say that Γ_0 can be continuously deformed into Γ_1.

Example 9.1. The circle $\{z \in \mathbb{C} : |z| = 1\}$ can be continuously deformed into the circle $\{z \in \mathbb{C} : |z| = 2\}$ in the complex plane $\mathbb{C}$ if both circles are traversed once in the counterclockwise direction.

Example 9.2. A unit square with center at the origin can be continuously deformed into the circle $\{z \in \mathbb{C} : |z| = 3\}$ in the complex plane $\mathbb{C}$ if both contours are traversed once in the counterclockwise direction.

Example 9.3. Let Γ_0 be a simple closed contour. (A simple contour is a contour that does not intersect itself except at the endpoints.) Suppose that the origin does not lie on Γ_0. Then Γ_0 can be continuously deformed into the origin in the complex plane $\mathbb{C}$. See Figure 9.1.

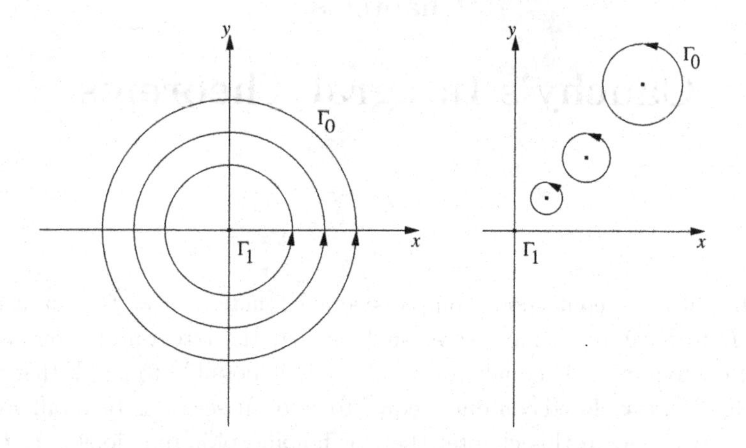

Fig. 9.1 Deformation of Γ_0 into the Origin

Example 9.4. Let $D = \{z \in \mathbb{C} : 1 < |z| < 4\}$. Let $\Gamma_0 = \{z \in \mathbb{C} : |z| = 3\}$ and let Γ_1 be a circle with center at 2 in the annulus $\{z \in \mathbb{C} : 1 < |z| < 3\}$. Then Γ_0 cannot be continuously deformed into Γ_1 in D. See Figure 9.2.

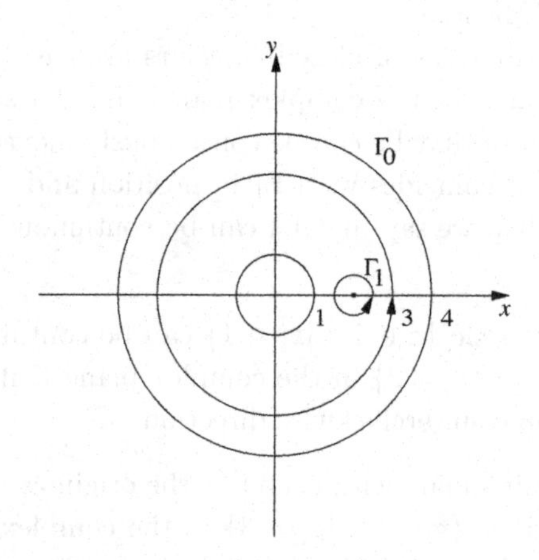

Fig. 9.2 Two Circles in an Annulus

The following theorem is a central result in complex analysis.

Theorem 9.5. *Let $w = f(z)$ be a holomorphic function on a domain D. Suppose that Γ_0 and Γ_1 are closed contours in D such that Γ_0 can be continuously deformed into Γ_1. Then*

$$\int_{\Gamma_0} f(z)\, dz = \int_{\Gamma_1} f(z)\, dz.$$

Remark 9.6. In most textbooks on complex analysis, a proof of the preceding theorem, also known as Cauchy's integral theorem, is given with the assumption that f' is continuous. As a matter of fact, the first proof without the assumption that f' is continuous is due to Edouard Goursat.

Definition 9.7. Let D be a domain in the complex plane such that every closed contour in D can be continuously deformed into a point in D. Then we say that D is a simply connected domain.

Example 9.8. Every disk is simply connected.

Example 9.9. The upper half plane $\{z \in \mathbb{C} : \operatorname{Im} z > 0\}$ is simply connected.

Example 9.10. Every annulus is not simply connected.

Here comes another version of Cauchy's integral theorem.

Theorem 9.11. *Let $w = f(z)$ be a holomorphic function on a simply connected domain D. Then for every closed contour Γ in D,*

$$\int_{\Gamma} f(z)\, dz = 0.$$

Theorem 9.11 follows from Theorem 9.5 because Γ can be continuously deformed into a point and the integral over a point is equal to zero.

Let us look at some instructive examples.

Example 9.12. Let Γ be the ellipse $\{x + iy \in \mathbb{C} : x^2 + 4y^2 = 1\}$ traversed once in the counterclockwise direction. Compute $\int_\Gamma \frac{1}{z} dz$.

Solution We see that Γ can be continuously deformed into a circle C with center at the origin and oriented once in the counterclockwise direction. Thus, by Cauchy's integral theorem and Example 8.4,

$$\int_\Gamma \frac{1}{z} dz = \int_C \frac{1}{z} dz = 2\pi i.$$

Example 9.13. Compute $\int_C \frac{e^z}{z^2-9}dz$, where C is the circle $\{z \in \mathbb{C} : |z| = 2\}$ traversed once in the counterclockwise direction.

Solution The function $\frac{e^z}{z^2-9}$ is holomorphic everywhere except at $z = \pm 3$. There is a simply connected neighborhood of the origin containing the circle $\{z \in \mathbb{C} : |z| = 2\}$ and excluding the singularities ± 3. By Cauchy's integral theorem, we get

$$\int_C \frac{e^z}{z^2-9}dz = 0.$$

Example 9.14. Let $a \in \mathbb{C}$. Compute $\int_\Gamma \frac{1}{z-a}dz$, where Γ is any circle not passing through the point a and traversed once in the counterclockwise direction.

Solution We look at two cases. The first case is when a is inside Γ. The second case is when a is outside Γ. For the first case, Γ can be continuously deformed into a circle C centered at a and oriented once in the counterclockwise direction. Thus, by Cauchy's integral theorem and Example 8.4,

$$\int_\Gamma \frac{1}{z-a}dz = \int_C \frac{1}{z-a}dz = 2\pi i.$$

For the second case, we can find an open disk containing Γ and excluding the point a. Thus, $\frac{1}{z-a}$ is holomorphic on the open disk. Thus, by Cauchy's integral theorem,

$$\int_\Gamma \frac{1}{z-a}dz = 0.$$

Example 9.15. Let Γ be a simple closed contour enclosing the point -1, excluding the points 0 and 1, and traversed once in the counterclockwise direction. See Figure 9.3. Compute $\int_\Gamma \frac{1}{z^2-1}dz$.

Solution Γ can be continuously deformed into a circle C with center at -1 and traversed once in the counterclockwise direction. Thus, by the preceding example,

$$\int_\Gamma \frac{1}{z^2-1}dz = \int_C \frac{1}{z^2-1}dz = \frac{1}{2}\int_C \frac{1}{z-1}dz - \frac{1}{2}\int_C \frac{1}{z+1}dz$$
$$= 0 - \frac{1}{2}2\pi i = -\pi i.$$

Appendix We give here a proof of Cauchy's integral theorem for a rectangle in a simply connected domain. The aim is to give some insight into the hypotheses on closed contours in simply connected domains. The proof

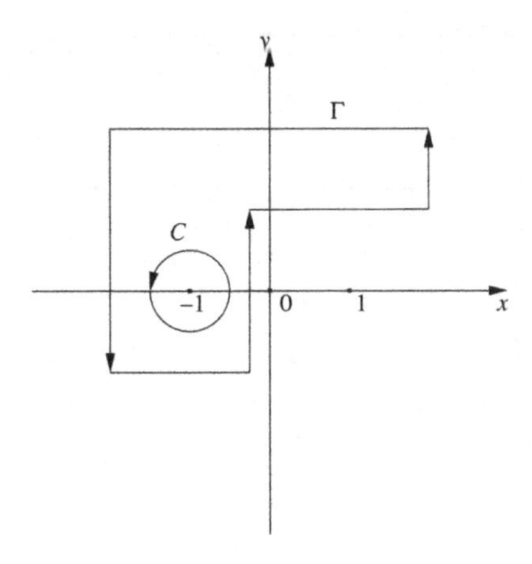

Fig. 9.3 The Contour Γ

is based on a property of the real numbers, which we now recall. This property can be found in most textbooks on real analysis.

Let $\{[a_n, b_n]\}_{n=1}^{\infty}$ be a sequence of closed intervals nested in the sense that

$$[a_{n+1}, b_{n+1}] \subseteq [a_n, b_n], \quad n = 1, 2, \ldots,$$

and

$$\lim_{n\to\infty}(b_n - a_n) = 0.$$

Then there exists a real number ξ such that

$$\xi \in \bigcap_{n=1}^{\infty}[a_n, b_n]$$

and

$$\lim_{n\to\infty} a_n = \lim_{n\to\infty} b_n = \xi.$$

We begin by dividing the rectangle into four rectangles as shown in Figure 9.4 and we denote the boundaries of the rectangles by Γ_1, Γ_2, Γ_3 and Γ_4. All boundaries are understood to be traversed once in the counterclockwise direction. Obviously, we get

$$\int_{\Gamma} f(z)\, dz = \sum_{j=1}^{4} \int_{\Gamma_j} f(z)\, dz$$

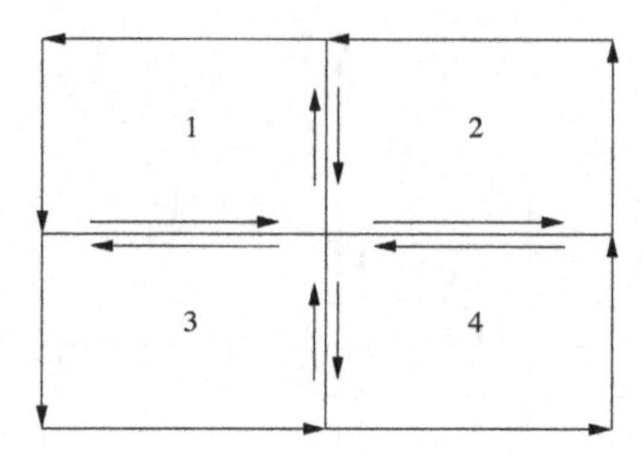

Fig. 9.4 The Rectangle in a Simply Connected Domain

and hence

$$\left| \int_{\Gamma} f(z)\, dz \right| \leq \sum_{j=1}^{4} \left| \int_{\Gamma_j} f(z)\, dz \right| .$$

One of the integrals on the right hand side of the above inequality has to be at least as large as the others and we denote it by $\left| \int_{C_1} f(z)\, dz \right|$. So,

$$\left| \int_{\Gamma} f(z)\, dz \right| \leq 4 \left| \int_{C_1} f(z)\, dz \right| .$$

Dividing the rectangle with boundary C_1 into four similar rectangles again as before, one of these four rectangles with boundary C_2 is such that

$$\left| \int_{C_1} f(z)\, dz \right| \leq 4 \left| \int_{C_2} f(z)\, dz \right| .$$

Therefore

$$\left| \int_{\Gamma} f(z)\, dz \right| \leq 4^2 \left| \int_{C_2} f(z)\, dz \right| .$$

Repeating this process gives a sequence $\{C_n\}_{n=1}^{\infty}$ of boundaries of rectangles $\{R_n\}_{n=1}^{\infty}$ such that

$$\left| \int_{\Gamma} f(z)\, dz \right| \leq 4^n \left| \int_{C_n} f(z)\, dz \right| .$$

The projections on the x-axis of the rectangles $\{R_n\}_{n=1}^{\infty}$ form a nested sequence of intervals where the length of each interval is half that of the preceding one. The same applies to the projections on the y-axis. So, we can find a point (ξ, η) such that the sequence $\{R_n\}_{n=1}^{\infty}$ converges to the complex number $\zeta = \xi + i\eta$ as $n \to \infty$. This means that for every positive number δ, there exists a positive integer N such that

$$n \geq N \Rightarrow T_n \subset \{z \in \mathbb{C} : |z - \zeta| < \delta\}.$$

Since $f'(\zeta)$ exists, we see that for every positive number ε, there exists a positive number δ such that

$$0 < |z - \zeta| < \delta \Rightarrow \left| \frac{f(z) - f(\zeta)}{z - \zeta} - f'(\zeta) \right| < \varepsilon.$$

So, for all z in $\mathbb{C}$ with $0 < |z - \zeta| < \delta$, we get

$$f(z) = f(\zeta) + (z - \zeta)f'(\zeta) + (z - \zeta)\omega(z),$$

where

$$|\omega(z)| < \varepsilon.$$

If we let N be such that

$$n \geq N \Rightarrow R_n \subset \{z \in \mathbb{C} : |z - \zeta| < \delta\},$$

then for $n \geq N$,

$$\int_{C_n} f(z)\,dz = f(\zeta) \int_{C_n} dz + f'(\zeta) \int_{C_n} (z - \zeta)\,dz + \int_{C_n} (z - \zeta)\omega(z)\,dz.$$

Since C_n is closed, it follows that

$$\int_{C_n} dz = \int_{C_n} (z - \zeta)\,dz = 0$$

and we get

$$\int_{C_n} f(z)\,dz = \int_{C_n} (z - \zeta)\omega(z)\,dz.$$

If we let L_n be the length of C_n, then

$$|z - \zeta| < L_n, \quad z \in R_n.$$

Thus, by Theorem 8.5, we get

$$\left| \int_{C_n} f(z)\,dz \right| \leq \varepsilon L_n^2, \quad n = 1, 2, \ldots.$$

But

$$L_n = 2^{-n} L,$$

where L is the length of Γ. Therefore

$$\left| \int_{\Gamma} f(z)\,dz \right| \leq 4^n \varepsilon L_n^2 = \varepsilon L^2.$$

Since ε is arbitrary, it follows that

$$\int_{C} f(z)\,dz = 0.$$

Exercises

(1) Does the function $w = f(z) = e^{z^2}$ have an antiderivative on $\mathbb{C}$? Explain your answer.

(2) Is $\int_\Gamma \overline{z}\, dz = 0$ for every closed contour Γ in $\mathbb{C}$? How do you reconcile your conclusion with Cauchy's integral theorem?

(3) Compute $\int_C \mathrm{Log}(z+3)\, dz$, where C is the circle with radius 2, centered at the origin and oriented once in the counterclockwise direction.

(4) Let Γ be the contour in Figure 9.5. Compute $\int_\Gamma \frac{2z^2-z+1}{(z-1)^2(z+1)} dz$.

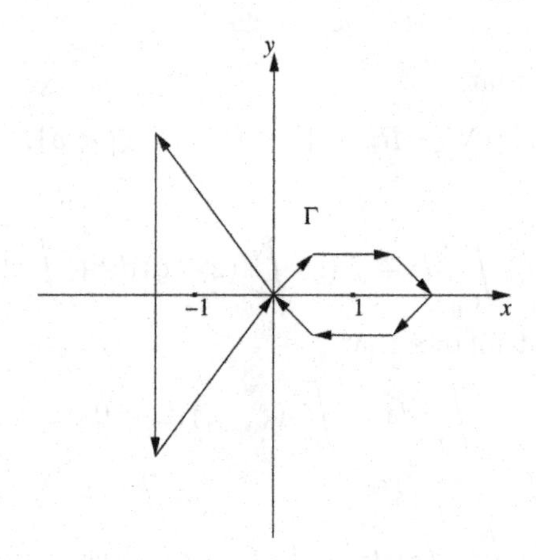

Fig. 9.5 The Contour Γ

Chapter 10

Cauchy's Integral Formulas

We give in this chapter a formula that is a landmark in complex analysis. It is known as Cauchy's integral formula. Let us emphasize the distinction between Cauchy's integral formula and Cauchy's integral theorem given in the last chapter. The latter is a result on the invariance of contour integrals, while the former, as we shall see in due course, allows us to compute values of a holomorphic function inside a contour in terms of the values of the function on the contour. It reveals the mean value property of holomorphic functions. See Exercise (3) in this chapter.

We state and prove Cauchy's integral formula for simple closed contours.

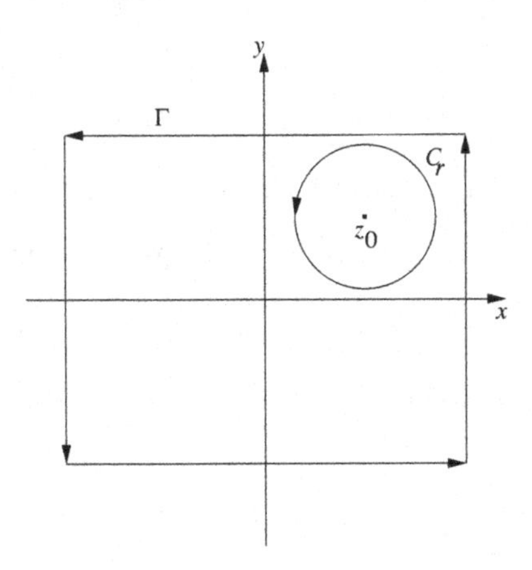

Fig. 10.1 Γ and C_r

Theorem 10.1. *Let $w = f(z)$ be a holomorphic function on a simply con-*

nected domain D. Let Γ be a simple closed contour in D oriented once in the counterclockwise direction. Then for every point z_0 inside Γ,

$$f(z_0) = \frac{1}{2\pi i} \int_\Gamma \frac{f(z)}{z - z_0} dz.$$

Proof We begin with the function $\frac{f(z)}{z-z_0}$. It is holomorphic everywhere on D except at z_0. Let C_r be the circle inside Γ with center at z_0 and radius r. We suppose that C_r is oriented once in the counterclockwise direction. Since D is simply connected, Γ can be continuously deformed into C_r. See Figure 10.1. Thus, by Cauchy's integral theorem,

$$\begin{aligned}\int_\Gamma \frac{f(z)}{z - z_0} dz &= \int_{C_r} \frac{f(z)}{z - z_0} dz \\ &= \int_{C_r} \frac{f(z_0)}{z - z_0} dz + \int_{C_r} \frac{f(z) - f(z_0)}{z - z_0} dz \\ &= f(z_0) \int_{C_r} \frac{1}{z - z_0} dz + \int_{C_r} \frac{f(z) - f(z_0)}{z - z_0} dz \\ &= f(z_0)\, 2\pi i + \int_{C_r} \frac{f(z) - f(z_0)}{z - z_0} dz.\end{aligned}$$

To estimate $\left|\int_{C_r} \frac{f(z)-f(z_0)}{z-z_0} dz\right|$, we note that on C_r,

$$\left|\frac{f(z) - f(z_0)}{z - z_0}\right| = \frac{|f(z) - f(z_0)|}{|z - z_0|} \le \frac{M_r}{r},$$

where

$$M_r = \max_{z \in C_r} |f(z) - f(z_0)|.$$

So, by the estimate in Theorem 8.5, we get

$$\left|\int_{C_r} \frac{f(z) - f(z_0)}{z - z_0} dz\right| \le \frac{M_r}{r} 2\pi r = 2\pi M_r.$$

Since f is continuous at z_0, it follows that $M_r \to 0$ ar $r \to 0+$. Therefore

$$\int_\Gamma \frac{f(z)}{z - z_0} dz = f(z_0) 2\pi i + \lim_{r \to 0+} \int_{C_r} \frac{f(z) - f(z_0)}{z - z_0} dz = f(z_0) 2\pi i.$$

Hence

$$f(z_0) = \frac{1}{2\pi i} \int_\Gamma \frac{f(z)}{z - z_0} dz,$$

as required. □

Example 10.2. Let Γ be the circle $\{z \in \mathbb{C} : |z - 2| = 3\}$ traversed once in the counterclockwise direction. Compute $\int_\Gamma \frac{e^z + \sin z}{z} dz$.

Solution Let $f(z) = e^z + \sin z$. Then f is holomorphic on and inside Γ. By Cauchy's integral formula, we get

$$\int_\Gamma \frac{e^z + \sin z}{z} dz = \int_\Gamma \frac{f(z)}{z-0} dz = 2\pi i f(0) = 2\pi i.$$

Example 10.3. Compute $\int_\Gamma \frac{\cos z}{z^2-4} dz$, where Γ is any simple closed contour in the right half plane enclosing the number 2 and oriented once in the counterclockwise direction.

Solution Let $f(z) = \frac{\cos z}{z+2}$. Then f is holomorphic on and inside Γ. By Cauchy's integral formula,

$$\int_\Gamma \frac{\cos z}{z^2-4} dz = \int_\Gamma \frac{f(z)}{z-2} dz = 2\pi i\, f(2) = \frac{2\pi i \cos 2}{4} = \frac{\pi i \cos 2}{2}.$$

Example 10.4. Compute $\int_C \frac{z^2 e^z}{2z+i} dz$, where C is the unit circle centered at the origin and traversed once in the clockwise direction.

Solution We write

$$\frac{z^2 e^z}{2z+i} = \frac{\frac{z^2 e^z}{2}}{z + \frac{i}{2}}$$

and we let

$$f(z) = \frac{z^2 e^z}{2}.$$

Then

$$\int_C \frac{z^2 e^z}{2z+i} dz = \int_C \frac{f(z)}{z+\frac{i}{2}} dz = -2\pi i\, f\left(-\frac{i}{2}\right) = \frac{\pi i}{4} e^{-i/2}.$$

To enhance the usefulness of Cauchy's integral formula, we need the following theorem.

Theorem 10.5. *Let Γ be a contour in $\mathbb{C}$. Let $w = g(z)$ be a continuous complex-valued function on Γ. Let G be the function defined on $\mathbb{C} - \Gamma$ by*

$$G(z) = \int_\Gamma \frac{g(\zeta)}{\zeta - z} d\zeta, \quad z \in \mathbb{C} - \Gamma.$$

Then G is holomorphic on $\mathbb{C} - \Gamma$ and

$$G'(z) = \int_\Gamma \frac{g(\zeta)}{(\zeta - z)^2} d\zeta, \quad z \in \mathbb{C} - \Gamma.$$

Proof Let $z \in \mathbb{C} - \Gamma$. We only need to prove that

$$\lim_{h\to 0} \frac{G(z+h)-G(z)}{h} = \int_\Gamma \frac{g(\zeta)}{(\zeta - z)^2} d\zeta.$$

But

$$\begin{aligned}
&\frac{G(z+h)-G(z)}{h} \\
&= \frac{1}{h}\int_\Gamma \left\{ \frac{1}{\zeta - (z+h)} - \frac{1}{\zeta - z} \right\} g(\zeta)\, d\zeta \\
&= \int_\Gamma \frac{g(\zeta)}{(\zeta - z - h)(\zeta - z)} d\zeta.
\end{aligned}$$

So,

$$\begin{aligned}
&\frac{G(z+h)-G(z)}{h} - \int_\Gamma \frac{g(\zeta)}{(\zeta - z)^2} d\zeta \\
&= \int_\Gamma \frac{g(\zeta)}{(\zeta - z - h)(\zeta - z)} d\zeta - \int_\Gamma \frac{g(\zeta)}{(\zeta - z)^2} d\zeta \\
&= h \int_\Gamma \frac{g(\zeta)}{(\zeta - z - h)(\zeta - z)^2} d\zeta = J_h(z),
\end{aligned}$$

where

$$J_h(z) = h \int_\Gamma \frac{g(\zeta)}{(\zeta - z - h)(\zeta - z)^2} d\zeta.$$

Now, we let d be the distance from z to Γ and we can assume that $|h| < \frac{d}{2}$. Then on Γ,

$$|\zeta - z - h| \geq |\zeta - z| - |h| \geq d - \frac{d}{2} = \frac{d}{2},$$

and hence

$$\left| \frac{g(\zeta)}{(\zeta - z - h)(\zeta - z)^2} \right| \leq \frac{M}{\frac{d}{2} d^2} = \frac{2M}{d^3},$$

where $M = \max_{\zeta \in \Gamma} |g(\zeta)|$. Using the estimate in Theorem 8.5, we get

$$|J_h(z)| \leq |\, h| \frac{2ML}{d^3},$$

where L is the length of Γ. So, $J_h(z) \to 0$ as $|h| \to 0$. Hence

$$\lim_{h\to 0} \frac{G(z+h)-G(z)}{h} = \int_\Gamma \frac{g(\zeta)}{(\zeta - z)^2} d\zeta,$$

as asserted. □

Remark 10.6. It can be proved similarly that for $z \subset D$,

$$G''(z) = 2\int_\Gamma \frac{g(\zeta)}{(\zeta - z)^3}d\zeta,\ G'''(z) = 6\int_\Gamma \frac{g(\zeta)}{(\zeta - z)^4}d\zeta, \ldots.$$

The general formula is

$$G^{(n)}(z) = n!\int_\Gamma \frac{g(\zeta)}{(\zeta - z)^{n+1}}d\zeta, \quad z \in D,$$

for $n = 1, 2, \ldots$.

Here is another remarkable result in complex analysis that has no analog in calculus.

Theorem 10.7. *Let $w = f(z)$ be a holomorphic function on a domain D. Then f', f'', ..., $f^{(n)}$, ... exist and are holomorphic on D.*

Proof Let $z_0 \in D$. Let C be a circle with center at z_0 such that D contains C and its inside. As usual, we suppose that C is oriented once in the counterclockwise direction. Now, for all z inside C, Cauchy's integral formula gives

$$f(z) = \frac{1}{2\pi i}\int_C \frac{f(\zeta)}{\zeta - z}d\zeta.$$

By Theorem 10.5,

$$f'(z) = \frac{1}{2\pi i}\int_C \frac{f(\zeta)}{(\zeta - z)^2}d\zeta$$

for all z inside C. Therefore by Remark 10.6,

$$f''(z) = \frac{2}{2\pi i}\int_C \frac{f(\zeta)}{(\zeta - z)^3}d\zeta$$

for all z inside C. Hence f' is differentiable on a neighborhood of z_0 and hence is holomorphic at z_0. But z_0 is arbitrary. Therefore f' is holomorphic on D. The same argument can be used for higher derivatives. □

By Cauchy's integral formula and Remark 10.6, we get the following result, which is also known as Cauchy's integral formula.

Theorem 10.8. *Let $w = f(z)$ be holomorphic on a simply connected domain D. Let Γ be a simple closed contour in D oriented once in the counterclockwise direction. Then for all z inside Γ,*

$$f^{(n)}(z) = \frac{n!}{2\pi i}\int_\Gamma \frac{f(\zeta)}{(\zeta - z)^{n+1}}d\zeta.$$

Example 10.9. Compute $\int_\Gamma \frac{e^{5z}}{z^3} dz$, where Γ is the unit circle with center at the origin and traversed once in the counterclockwise direction.

Solution z^3 suggests using the second derivative of $f(z) = e^{5z}$. So,

$$f''(0) = \frac{2}{2\pi i} \int_\Gamma \frac{e^{5z}}{z^3} dz.$$

Therefore

$$\int_\Gamma \frac{e^{5z}}{z^3} dz = \frac{1}{2} 2\pi i \, 25 \, e^0 = 25\pi i.$$

Example 10.10. Let Γ be the contour in Figure 10.2. Compute $\int_\Gamma \frac{2z+1}{z(z-1)^2} dz$.

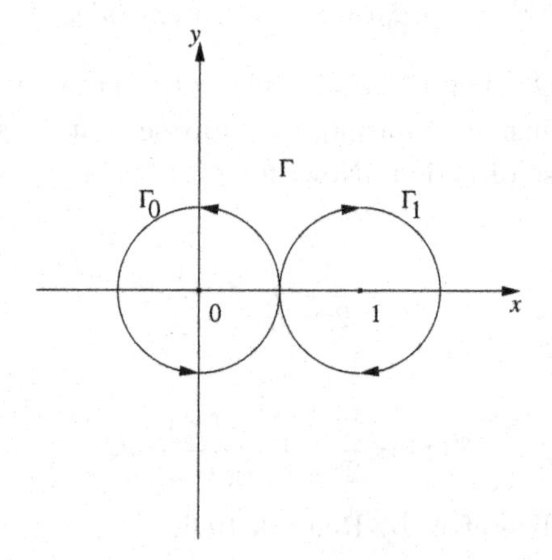

Fig. 10.2 The ∞-Contour

Solution Obviously,

$$\int_\Gamma \frac{2z+1}{z(z-1)^2} dz = \int_{\Gamma_0} \frac{2z+1}{z(z-1)^2} dz + \int_{\Gamma_1} \frac{2z+1}{z(z-1)^2} dz.$$

By Cauchy's integral formula,

$$\int_{\Gamma_0} \frac{2z+1}{z(z-1)^2} dz = 2\pi i \left. \frac{2z+1}{(z-1)^2} \right|_{z=0} = 2\pi i.$$

By Cauchy's integral formula for the first derivative,

$$\int_{\Gamma_1} \frac{2z+1}{z(z-1)^2} dz = -2\pi i \left. \frac{d}{dz} \left(\frac{2z+1}{z} \right) \right|_{z=1} = 2\pi i \left. \left(\frac{1}{z^2} \right) \right|_{z=1} = 2\pi i.$$

Thus,

$$\int_{\Gamma} \frac{2z+1}{z(z-1)^2} dz = 4\pi i.$$

Example 10.11. Let Γ be the contour self-intersecting at 2. See Figure 10.3. Compute $\int_{\Gamma} \frac{\cos z}{z^2(z-3)} dz$.

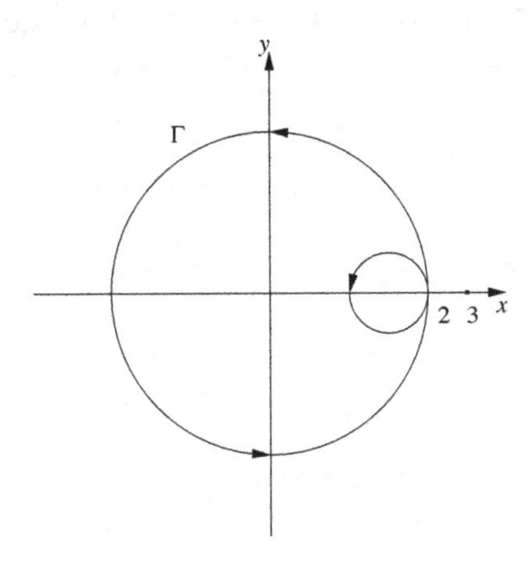

Fig. 10.3 A Self-Intersecting Contour

Solution Let Γ_b be the big circle and Γ_s be the small circle. By Cauchy's integral theorem,

$$\int_{\Gamma_s} \frac{\cos z}{z^2(z-3)} dz = 0.$$

By Cauchy's integral formula,

$$\int_{\Gamma_b} \frac{\cos z}{z^2(z-3)} dz = 2\pi i f'(0),$$

where

$$f(z) = \frac{\cos z}{z-3}.$$

But

$$f'(z) = \frac{-(z-3)\sin z - \cos z}{(z-3)^2}$$

and hence

$$\int_{\Gamma} \frac{\cos z}{z^2(z-3)} dz = \int_{\Gamma_b} \frac{\cos z}{z^2(z-3)} dz = -\frac{2}{9}\pi i.$$

Exercises

(1) Compute $\int_C \frac{\cos z}{z^{28}} dz$ and $\int_C \left(\frac{z-2}{2z-1}\right)^3 dz$ where C is the unit circle with center at the origin and oriented once in the counterclockwise direction.

(2) Compute $\int_\Gamma \frac{e^{iz}}{(z^2+1)^2} dz$, where Γ is the circle $\{z \in \mathbb{C} : |z| = 3\}$ oriented once in the counterclockwise direction.

(3) **(Mean Value Property of Holomorphic Functions)** Suppose that $w = f(z)$ is holomorphic on and inside the circle $\{z \in \mathbb{C} : |z - z_0| = r\}$. Prove that

$$f(z_0) = \frac{1}{2\pi} \int_0^{2\pi} f(z_0 + re^{i\theta})\, d\theta.$$

(4) Under the hypotheses of the preceding exercise, prove that for $n = 0, 1, 2, \ldots$,

$$f^{(n)}(z_0) = \frac{n!}{2\pi r^n} \int_0^{2\pi} f(z_0 + re^{i\theta}) e^{-in\theta} d\theta.$$

(5) Let $w = f(z)$ be a holomorphic function on and inside the unit circle with center at the origin. Prove that

$$\int_{|z| \le 1} f(x + iy)\, dx\, dy = \pi f(0).$$

(6) **(Cauchy's Estimates)** Let $w = f(z)$ be a holomorphic function on and inside the circle C_R given by

$$C_R = \{z \in \mathbb{C} : |z - z_0| = R\}.$$

Prove that if

$$|f(z)| \le M, \quad z \in C_R,$$

then

$$|f^{(n)}(z_0)| \le \frac{Mn!}{R^n}, \quad n = 0, 1, 2, \ldots.$$

(7) **(Liouville's Theorem)** Prove that every bounded entire function must be a constant function.

(8) Let f be an entire function such that there exists a positive number M for which

$$\operatorname{Re} f(z) \le M, \quad z \in \mathbb{C}.$$

Prove that f has to be a constant function.

(9) Let u be a real-valued function on a domain D of the complex plane $\mathbb{C}$ such that $u \in C^2(D)$ and

$$\frac{\partial^2 u}{\partial x^2} + \frac{\partial^2 u}{\partial y^2} = 0$$

for all $z = x + iy \in \mathbb{D}$. Then u is said to be harmonic on D. Prove that the real and imaginary parts of a holomorphic function on a domain D of the complex plane $\mathbb{C}$ are harmonic on D. ($u \in C^2(D)$ means that u and all its partial derivatives up to and including the second order are continuous on D.)

(10) Prove Liouville's theorem for harmonic functions on $\mathbb{C}$ to the effect that every bounded harmonic function on the whole complex plane has to be a constant function.

(11) **(Fundamental Theorem of Algebra)** Let P be a polynomial given by

$$P(z) = a_0 + a_1 z + a_2 z^2 + \cdots + a_n z^n, \quad z \in \mathbb{C},$$

where $a_0, a_1, a_2, \ldots, a_n$ are complex constants and n is a positive integer. Prove that P has a zero in $\mathbb{C}$ in the sense that there exists a complex number z_0 such that $P(z_0) = 0$.

Chapter 11

Taylor Series and Power Series

The first goal of this chapter is to prove that holomorphic functions can be locally represented by Taylor series.

Definition 11.1. Let $w = f(z)$ be a holomorphic function at z_0. Then the series $\sum_{n=0}^{\infty} \frac{f^{(n)}(z_0)}{n!}(z - z_0)^n$ is called the Taylor series of f at z_0. In the special case when $z_0 = 0$, we call it the Maclaurin series of f.

We can always write down the Taylor series. The real issue is whether or not it converges to the function in question.

Theorem 11.2. *Let $w = f(z)$ be a holomorphic function on the open disk $\{z \in \mathbb{C} : |z - z_0| < R\}$. Then the Taylor series $\sum_{n=0}^{\infty} \frac{f^{(n)}(z_0)}{n!}(z - z_0)^n$ converges to $f(z)$ uniformly on every closed subdisk $\{z \in \mathbb{C} : |z - z_0| \leq \rho\}$.*

Proof Let C be the circle $\{z \in \mathbb{C} : |z - z_0| = r\}$ oriented once in the counterclockwise direction, where $\rho < r < R$. Then for all z in the closed subdisk $\{z \in \mathbb{C} : |z - z_0| \leq \rho\}$, Cauchy's integral formula gives

$$f(z) = \frac{1}{2\pi i} \int_C \frac{f(\zeta)}{\zeta - z} d\zeta.$$

Now, for every number $\omega \neq 1$, we get

$$1 + \omega + \omega^2 + \cdots + \omega^n = \frac{1 - \omega^{n+1}}{1 - \omega}$$

and hence

$$1 + \omega + \omega^2 + \cdots + \omega^n + \frac{\omega^{n+1}}{1 - \omega} = \frac{1}{1 - \omega}.$$

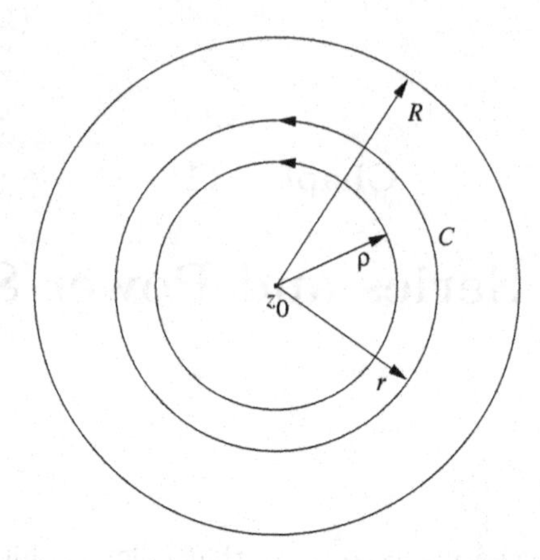

Fig. 11.1 Uniform Convergence on Subdisks

Therefore

$$\frac{1}{\zeta - z} = \frac{1}{(\zeta - z_0) - (z - z_0)} = \frac{1}{1 - \frac{z-z_0}{\zeta - z_0}} \frac{1}{\zeta - z_0}$$

$$= \left[1 + \frac{z - z_0}{\zeta - z_0} + \frac{(z-z_0)^2}{(\zeta - z_0)^2} + \cdots + \frac{(z-z_0)^n}{(\zeta - z_0)^n} + \frac{\frac{(z-z_0)^{n+1}}{(\zeta - z_0)^{n+1}}}{1 - \frac{z-z_0}{\zeta - z_0}}\right] \frac{1}{\zeta - z_0}$$

$$= \sum_{j=0}^{n} \frac{(z-z_0)^j}{(\zeta - z_0)^{j+1}} + \frac{(z-z_0)^{n+1}}{(\zeta - z_0)^{n+1}} \frac{1}{\zeta - z}.$$

So,

$$f(z) = \frac{1}{2\pi i} \sum_{j=0}^{n} (z - z_0)^j \int_C \frac{f(\zeta)}{(\zeta - z_0)^{j+1}} d\zeta + T_n(z),$$

where

$$T_n(z) = \frac{1}{2\pi i} \int_C \frac{f(\zeta)}{\zeta - z} \frac{(z-z_0)^{n+1}}{(\zeta - z_0)^{n+1}} d\zeta.$$

For $j = 0, 1, 2, \ldots,$ Cauchy's integral formula gives

$$\frac{1}{2\pi i} \int_C \frac{f(\zeta)}{(\zeta - z_0)^{j+1}} d\zeta = \frac{f^{(j)}(z_0)}{j!}.$$

Now, for $\zeta \in C$, we get

$$\frac{|z - z_0|}{|\zeta - z_0|} \leq \frac{\rho}{r}$$

and

$$|\zeta - z| \geq r - \rho.$$

So,

$$|T_n(z)| \leq \frac{1}{2\pi} \max_{\zeta \in C} |f(\zeta)| \left(\frac{\rho}{r}\right)^{n+1} \frac{1}{r-\rho} 2\pi r \to 0$$

uniformly on $\{z \in \mathbb{C} : |z - z_0| \leq \rho\}$ as $n \to \infty$. Therefore

$$f(z) = \sum_{n=0}^{\infty} \frac{f^{(n)}(z_0)}{n!}(z - z_0)^n$$

uniformly on $\{z \in \mathbb{C} : |z - z_0| \leq \rho\}$. □

Example 11.3. Find the Taylor series of $\operatorname{Log} z$ at $z_0 = 1$. State also the largest disk of convergence centered at 1.

Solution Let $f(z) = \operatorname{Log} z$. Then f is holomorphic at 1 and

$$\{z \in \mathbb{C} : |z - 1| < 1\}$$

is the largest disk centered at 1 on which f is holomorphic. Now,

$$f'(z) = \frac{1}{z} \Rightarrow f'(1) = 1,$$

$$f''(z) = -\frac{1}{z^2} \Rightarrow f''(1) = -1,$$

$$f'''(z) = \frac{2}{z^3} \Rightarrow f'''(1) = 2,$$

$$\cdots.$$

Thus,

$$\begin{aligned}
\operatorname{Log} z &= f(1) + f'(1)(z-1) + \frac{f''(1)}{2!}(z-1)^2 + \frac{f'''(1)}{3!}(z-1)^3 + \cdots \\
&= (z-1) - \frac{1}{2}(z-1)^2 + \frac{1}{3}(z-1)^3 - \cdots \\
&= \sum_{n=1}^{\infty} \frac{(-1)^{n+1}}{n}(z-1)^n, \quad |z-1| < 1.
\end{aligned}$$

Example 11.4. Find the Taylor series of $\frac{1}{1-z}$ at $z_0 = 0$. State also the largest disk of convergence centered at 0.

Solution Let $f(z) = \frac{1}{1-z}$. Then f is holomorphic at 0 and

$$\{z \in \mathbb{C} : |z| < 1\}$$

is the largest disk centered at 0 on which f is holomorphic. Now,

$$f'(z) = \frac{1}{(1-z)^2} \Rightarrow f'(0) = 1,$$

$$f''(z) = \frac{2}{(1-z)^3} \Rightarrow f''(0) = 2,$$

$$f'''(z) = \frac{3!}{(1-z)^4} \Rightarrow f'''(0) = 3!,$$

$$\cdots.$$

So,

$$\begin{aligned}\frac{1}{1-z} &= f(0) + f'(0)z + \frac{f''(0)}{2!}z^2 + \cdots \\ &= 1 + z + z^2 + z^3 + \cdots \\ &= \sum_{n=0}^{\infty} z^n, \quad |z| < 1,\end{aligned}$$

which is the well-known geometric series.

Example 11.5. Find the Taylor series of e^z at $z_0 = 0$. State the largest disk of convergence centered at 0.

Solution Let $f(z) = e^z$. Then f is entire and the whole complex plane is the disk on which f is holomorphic. Obviously, for $n = 0, 1, 2, \ldots$,

$$f^{(n)}(z) = e^z \Rightarrow f^{(n)}(0) = 1.$$

Thus,

$$e^z = \sum_{n=0}^{\infty} \frac{f^{(n)}(0)}{n!} z^n = \sum_{n=0}^{\infty} \frac{z^n}{n!}, \quad z \in \mathbb{C}.$$

Question If $w = f(z)$ is holomorphic at z_0, then so is f'. Is the Taylor series of f' related to that of f?

Theorem 11.6. *Let $w = f(z)$ be holomorphic at z_0. Then the Taylor series of f' at z_0 can be obtained from that of f at z_0 by differentiating term by term, and it converges on the same disk as that of f.*

Proof Suppose that

$$f(z) = f(z_0) + f'(z_0)(z - z_0) + \frac{f''(z_0)}{2!}(z - z_0)^2 + \cdots, \quad |z - z_0| < R,$$

where $\{z \in \mathbb{C} : |z - z_0| < R\}$ is the largest disk on which f is holomorphic. Differentiating the series term by term, we get

$$f'(z_0) + f''(z_0)(z - z_0) + \frac{f'''(z_0)}{2!}(z - z_0)^2 + \cdots,$$

which is the Taylor series of f' at z_0. Since f' is holomorphic on the disk $\{z \in \mathbb{C} : |z - z_0| < R\}$, it follows that the Taylor series of f' converges on $\{z \in \mathbb{C} : |z - z_0| < R\}$. To see that this is the largest disk on which the Taylor series of f' converges, suppose that the Taylor series of f' converges on a larger disk $\{z \in \mathbb{C} : |z - z_0| < R'\}$. Because of uniform convergence, we can integrate term by term and the resulting series

$$f'(z_0)(z - z_0) + \frac{f''(z_0)}{2!}(z - z_0)^2 + \frac{f'''(z_0)}{3!}(z - z_0)^3 + \cdots$$

converges on $\{z \in \mathbb{C} : |z - z_0| < R'\}$ and this is a contradiction. □

Example 11.7. Find the Maclaurin series of $\sin z$ and $\cos z$.

Solution Let $f(z) = \sin z$. Then

$$f'(z) = \cos z \Rightarrow f'(0) = 1,$$

$$f''(z) = -\sin z \Rightarrow f''(0) = 0,$$

$$f'''(z) = -\cos z \Rightarrow f'''(0) = -1,$$

$$f^{(4)}(z) = \sin z \Rightarrow f^{(4)}(0) = 0,$$

$$\cdots.$$

So,

$$\sin z = z - \frac{z^3}{3!} + \frac{z^5}{5!} - \frac{z^7}{7!} + \cdots, \quad z \in \mathbb{C}.$$

By Theorem 11.6,

$$\cos z = 1 - \frac{z^2}{2!} + \frac{z^4}{4!} + \frac{z^6}{6!} - \cdots, \quad z \in \mathbb{C}.$$

The following example illustrates the point that sometimes we need to use an indirect method to compute a Taylor series or Maclaurin series.

Example 11.8. Find the first three terms of the Maclaurin series for $\tan z$.

Solution First let us see where $\tan z$ is holomorphic. Since

$$\tan z = \frac{\sin z}{\cos z},$$

we see that $\tan z$ is holomorphic at all points z unless $\cos z = 0$. But

$$\cos z = 0 \Leftrightarrow e^{iz} = -e^{-iz} \Leftrightarrow e^{2iz-\pi i} = 1 \Leftrightarrow 2iz - \pi i = 2k\pi i,$$

where $k \in \mathbb{Z}$. Therefore $\tan z$ is holomorphic everywhere except at

$$z = \left(k + \frac{1}{2}\right)\pi, \quad k \in \mathbb{Z}.$$

Therefore $\{z \in \mathbb{C} : |z| < \frac{\pi}{2}\}$ is the largest disk on which $\tan z$ is holomorphic. Now, suppose that the Maclaurin series of $\tan z$ is given by

$$\tan z = \sum_{n=0}^{\infty} a_n z^n, \quad |z| < \frac{\pi}{2}.$$

Since $\tan z = \frac{\sin z}{\cos z}$, we get

$$\cos z \tan z = \sin z$$

and hence

$$\left(1 - \frac{z^2}{2!} + \frac{z^4}{4!} - \cdots\right)(a_0 + a_1 z + a_2 z^2 + a_3 z^3 + a_4 z^4 + a_5 z^5 + \cdots)$$
$$= a_0 + a_1 z + \left(a_2 - \frac{a_0}{2!}\right) z^2 + \left(a_3 - \frac{a_1}{2!}\right) z^3$$
$$+ \left(a_4 - \frac{a_2}{2!} + \frac{a_0}{4!}\right) z^4 + \left(a_5 - \frac{a_3}{2!} + \frac{a_1}{4!}\right) z^5 + \cdots$$

and this is the same as

$$z - \frac{z^3}{3!} + \frac{z^5}{5!} - \cdots.$$

Therefore

$$a_0 = 0,\ a_1 = 1,\ a_2 = 0,\ a_3 = \frac{1}{3},\ a_4 = 0,\ a_5 = \frac{2}{15}, \ldots.$$

Hence

$$\tan z = z + \frac{1}{3}z^3 + \frac{2}{15}z^5 + \cdots, \quad |z| < \frac{\pi}{2}.$$

We now look at series that are apparently more general than Taylor series. A series of the form $\sum_{n=0}^{\infty} a_n (z - z_0)^n$ is called a power series at z_0. Obviously, it converges at z_0. The most basic property of power series is given in the following theorem.

Theorem 11.9. *Let $\sum_{n=0}^{\infty} a_n(z-z_0)^n$ be a power series. Then there exists a real number R in $[0,\infty]$ such that*

(1) the series converges absolutely on $\{z \in \mathbb{C} : |z-z_0| < R\}$,
(2) the series diverges on $\{z \in \mathbb{C} : |z-z_0| > R\}$,
(3) the series converges uniformly on every closed subdisk of the disk of convergence $\{z \in \mathbb{C} : |z-z_0| < R\}$.

In fact,

$$R = \frac{1}{\limsup_{n\to\infty} \sqrt[n]{|a_n|}}.$$

Remark 11.10. We call R the radius of convergence of the power series. The formula for R in Theorem 11.9 is known as Hadamard's radius of convergence. If $R=\infty$, then the power series converges at every complex number z.

Proof of Theorem 11.9 Without loss of generality, we assume that $z_0 = 0$. Suppose that $R=0$. Then

$$\limsup_{n\to\infty} \sqrt[n]{|a_n|} = \infty.$$

Let z be a nonzero complex number. Then

$$\limsup_{n\to\infty} \sqrt[n]{|a_n z^n|} = |z| \limsup_{n\to\infty} \sqrt[n]{|a_n|} = \infty.$$

So, by Cauchy's root test, the power series $\sum_{n=0}^{\infty} a_n z^n$ diverges for all nonzero complex numbers z. Next, we suppose that $R=\infty$. Then

$$\limsup_{n\to\infty} \sqrt[n]{|a_n|} = 0.$$

Let $z \in \mathbb{C}$. Then

$$\limsup_{n\to\infty} \sqrt[n]{|a_n z^n|} = |z| \limsup_{n\to\infty} \sqrt[n]{|a_n|} = 0.$$

So, by Cauchy's root test, the power series $\sum_{n=0}^{\infty} a_n z^n$ converges absolutely for all z in $\mathbb{C}$. Let $\rho \in [0,\infty)$. Then for all z in the disk $\{z \in \mathbb{C} : |z| \le \rho\}$,

$$|a_n z^n| \le |a_n|\rho^n, \quad n = 0,1,2,\ldots.$$

Since $\sum_{n=0}^{\infty} |a_n|\rho^n < \infty$, it follows from Weierstrass' M-test that $\sum_{n=0}^{\infty} a_n z^n$ converges uniformly on $\{z \in \mathbb{C} : |z| \le \rho\}$. Finally, we suppose that $0 < R < \infty$. Then

$$\limsup_{n\to\infty} \sqrt[n]{|a_n|} = \frac{1}{R}.$$

Let $z \in \mathbb{C}$. Then

$$\limsup_{n\to\infty} \sqrt[n]{a_n z^n} = |z| \limsup_{n\to\infty} \sqrt[n]{|a_n|} \begin{cases} < 1, & |z| < R, \\ > 1, & |z| > R. \end{cases}$$

So, by Cauchy's root test, the power series $\sum_{n=0}^{\infty} a_n z^n$ converges absolutely on $\{z \in \mathbb{C} : |z| < R\}$ and diverges on $\{z \in \mathbb{C} : |z| > R\}$. That the convergence is uniform on every closed subdisk is the same as before.

The most striking property of power series is given in the following theorem.

Theorem 11.11. *Suppose that*

$$f(z) = \sum_{n=0}^{\infty} a_n (z - z_0)^n, \quad |z - z_0| < R,$$

where R is the radius of convergence of the power series. Then f is holomorphic on the disk $\{z \in \mathbb{C} : |z - z_0| < R\}$.

Theorem 11.11 explains why the open disk is one of the most important simply connected domains in complex analysis. The proof of Theorem 11.11 depends on the following result, which is known as Weierstrass' theorem.

Theorem 11.12. *Let $\{f_n\}_{n=1}^{\infty}$ be a sequence of holomorphic functions on a simply connected domain D such that $f_n \to f$ uniformly on D as $n \to \infty$. Then f is holomorphic on D.*

Proof Since f is the uniform limit of a sequence of continuous functions on D, f is continuous on D. Let Γ be any closed contour in D. Then, by uniform convergence again,

$$\int_\Gamma f(z)\, dz = \lim_{n \to \infty} \int_\Gamma f_n(z)\, dz.$$

By Cauchy's integral theorem,

$$\int_\Gamma f_n(z)\, dz = 0, \quad n = 1, 2, \ldots,$$

and hence

$$\int_\Gamma f(z)\, dz = 0.$$

So, by Theorem 8.11, f has an antiderivative F on D, *i.e.*,

$$F'(z) = f(z), \quad z \in D.$$

Therefore F is holomorphic on D. By Theorem 10.7, all derivatives of F are then holomorphic on D. Hence f is holomorphic on D. □

Proof of Theorem 11.11 Let $z_1 \in \mathbb{C}$ be such that $|z_1 - z_0| < R$. Let ρ be a real number such that $|z_1 - z_0| < \rho < R$. Then the power series converges uniformly to f on the disk $\{z \in \mathbb{C} : |z - z_0| \leq \rho\}$. Therefore, by Theorem 11.12, f is holomorphic on $\{z \in \mathbb{C} : |z - z_0| < \rho\}$ and hence at z_1.

Once we have Theorem 11.11, we can use Taylor's theorem, *i.e.*, Theorem 11.2, to get the following theorem.

Theorem 11.13. *Suppose that*

$$f(z) = \sum_{n=0}^{\infty} a_n (z - z_0)^n, \quad |z - z_0| < R,$$

where $R > 0$. *Then*

$$a_n = \frac{f^{(n)}(z_0)}{n!}, \quad n = 0, 1, 2, \ldots.$$

Thus, the power series is the Taylor series of f *at* z_0.

Remark 11.14. It is interesting to note that a power series with positive radius of convergence is a holomorphic function inside the radius of convergence. Since a holomorphic function has a Taylor series expansion, it turns out that a power series with positive radius of convergence is a Taylor series.

Exercises

(1) Use the Taylor series of $\operatorname{Log} z$ to find the Taylor series of $\frac{1}{z}$ at 1. What is the largest disk of convergence centered at 1?
(2) Find the Maclaurin series of e^{z^2}.
(3) Find the radius of convergence of the power series $\sum_{n=1}^{\infty} n^2 z^n$. Find the holomorphic function with power series $\sum_{n=1}^{\infty} n^2 z^n$.
(4) Find the radius of convergence of the power series $\sum_{n=0}^{\infty} a_n z^n$, where

$$a_n = \begin{cases} 2^n, & n = 0, 2, 4, \ldots, \\ 1, & n = 1, 3, 5, \ldots. \end{cases}$$

(5) Find the holomorphic function with power series given in the preceding exercise.

Chapter 12

Laurent Series and Isolated Singularities

Up to this point, we know that a holomorphic function on an open disk can be represented by a power series. What happens if a function is holomorphic only on an annulus?

Theorem 12.1. *Let $w = f(z)$ be a holomorphic function on an annulus $D = \{z \in \mathbb{C} : r < |z - z_0| < R\}$. Then*

$$f(z) = \sum_{n=0}^{\infty} a_n(z - z_0)^n + \sum_{n=1}^{\infty} a_{-n}(z - z_0)^{-n},$$

where both series converge on the annulus and converge uniformly on every closed subannulus $\{z \in \mathbb{C} : \rho_1 \leq |z - z_0| \leq \rho_2\}$ of D. Moreover,

$$a_n = \frac{1}{2\pi i} \int_C \frac{f(\zeta)}{(\zeta - z_0)^{n+1}} d\zeta, \quad n = 0, \pm 1, \pm 2, \ldots,$$

where C is any simple closed contour in the annulus D enclosing z_0 and oriented once in the counterclockwise direction.

Remark 12.2. We allow $r = 0$ or $R = \infty$. The coefficient a_{-1} of $(z - z_0)^{-1}$ is a significant number in complex analysis. It is known as the residue of the function $w = f(z)$ at z_0.

We call $\sum_{n=0}^{\infty} a_n(z - z_0)^n + \sum_{n=1}^{\infty} a_{-n}(z - z_0)^{-n}$ the Laurent series of f at z_0. The series $\sum_{n=1}^{\infty} a_{-n}(z - z_0)^{-n}$ is called the singular part of f at z_0. If f is holomorphic on the whole disk $\{z \in \mathbb{C} : |z - z_0| < R\}$, then the singular part disappears.

Proof of Theorem 12.1 Let us first prove uniform convergence to $f(z)$ on the closed subannulus $A = \{z \in \mathbb{C} : \rho_1 \leq |z - z_0| \leq \rho_2\}$. Let $z \in A$. Let C_1 and C_2 be circles in A centered at z_0 such that z is outside C_1 and inside C_2. Let R_1 be the radius of C_1 and let R_2 be the radius of C_2. The inner

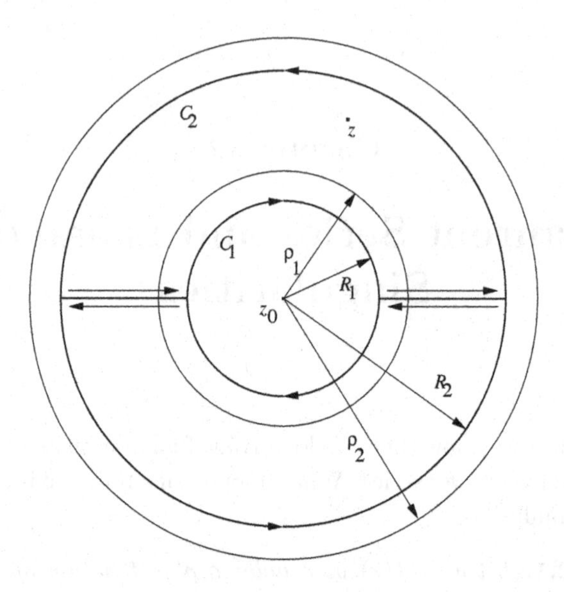

Fig. 12.1 Uniform Convergence on Subannuli

circle C_1 is oriented once in the clockwise direction and the outer circle C_2 is oriented once in the counterclockwise direction. See Figure 12.1. Let Γ_+ be the simple closed contour consisting of the upper semicircle of C_2, the two line segments from left to right, and the upper semicircle of C_1. Let Γ_- be the analog of Γ_+ with upper replaced by lower, and left and right interchanged. Both Γ_+ and Γ_- are oriented once in the counterclockwise direction. Using Cauchy's integral formula for $f(\zeta)$ on Γ_+ and Cauchy's integral theorem for $\frac{f(\zeta)}{\zeta - z}$ on Γ_-, we get

$$\begin{aligned} f(z) &= \frac{1}{2\pi i}\int_{\Gamma_+}\frac{f(\zeta)}{\zeta - z}d\zeta + \frac{1}{2\pi i}\int_{\Gamma_-}\frac{f(\zeta)}{\zeta - z}d\zeta \\ &= \frac{1}{2\pi i}\int_{C_1}\frac{f(\zeta)}{\zeta - z}d\zeta + \frac{1}{2\pi i}\int_{C_2}\frac{f(\zeta)}{\zeta - z}d\zeta. \end{aligned}$$

Since z is inside C_2, the proof of Theorem 11.2 on the Taylor expansion can be used and we get

$$\frac{1}{2\pi i}\int_{C_2}\frac{f(\zeta)}{\zeta - z}d\zeta = \sum_{n=0}^{\infty} a_n(z - z_0)^n,$$

where the convergence is uniform on $\{z \in \mathbb{C} : |z - z_0| \le \rho_2\}$ and

$$a_n = \frac{1}{2\pi i}\int_{C_2}\frac{f(\zeta)}{(\zeta - z_0)^{n+1}}d\zeta, \quad n = 0, 1, 2, \ldots.$$

(Note that for $n = 0, 1, 2, \ldots$, we cannot conclude that

$$a_n = \frac{f^{(n)}(z_0)}{n!}$$

because f is holomorphic only on an annulus centered at z_0.) Now, for $\frac{1}{2\pi i}\int_{C_1}\frac{f(\zeta)}{\zeta - z}d\zeta$, we write

$$\begin{aligned}
&\frac{1}{\zeta - z}\\
&= \frac{1}{(\zeta - z_0) - (z - z_0)} = -\frac{1}{z - z_0}\frac{1}{1 - \frac{\zeta - z_0}{z - z_0}}\\
&= -\frac{1}{z - z_0}\left[1 + \frac{\zeta - z_0}{z - z_0} + \frac{(\zeta - z_0)^2}{(z - z_0)^2} + \cdots + \frac{(\zeta - z_0)^m}{(z - z_0)^m} + \frac{\frac{(\zeta - z_0)^{m+1}}{(z - z_0)^{m+1}}}{1 - \frac{\zeta - z_0}{z - z_0}}\right]\\
&= -\left[\frac{1}{z - z_0} + \frac{\zeta - z_0}{(z - z_0)^2} + \cdots + \frac{(\zeta - z_0)^m}{(z - z_0)^{m+1}}\right] - \frac{1}{z - \zeta}\frac{(\zeta - z_0)^{m+1}}{(z - z_0)^{m+1}}.
\end{aligned}$$

So,

$$\frac{1}{2\pi i}\int_{C_1}\frac{f(\zeta)}{\zeta - z}d\zeta = \sum_{j=1}^{m+1} a_{-j}(z - z_0)^{-j} + R_m(z),$$

where

$$a_{-j} = -\frac{1}{2\pi i}\int_{C_1}\frac{f(\zeta)}{(\zeta - z_0)^{-j+1}}d\zeta, \quad j = 1, 2, \ldots, m + 1,$$

and

$$R_m(z) = \frac{1}{2\pi i}\int_{C_1}\frac{f(\zeta)}{\zeta - z}\frac{(\zeta - z_0)^{m+1}}{(z - z_0)^{m+1}}d\zeta.$$

To estimate $|R_m(z)|$, we note that on C_1,

$$|\zeta - z| \geq |z - z_0| - |\zeta - z_0| \geq \rho_1 - R_1$$

and

$$\frac{|\zeta - z_0|^{m+1}}{|z - z_0|^{m+1}} \leq \left(\frac{R_1}{\rho_1}\right)^{m+1}.$$

Therefore

$$|R_m(z)| \leq \frac{1}{2\pi}\max_{\zeta \in C_1}|f(\zeta)|\frac{1}{\rho_1 - R_1}\left(\frac{R_1}{\rho_1}\right)^{m+1} \to 0$$

as $m \to \infty$. Thus,

$$\frac{1}{2\pi i}\int_{C_1}\frac{f(\zeta)}{\zeta - z}d\zeta = \sum_{n=1}^{\infty} a_{-n}(z - z_0)^{-n}$$

and the convergence is uniform on A. It remains to prove that

$$a_n = \frac{1}{2\pi i}\int_C \frac{f(\zeta)}{(\zeta - z)^{n+1}}d\zeta, \quad n = 0, \pm 1, \pm 2, \ldots,$$

where C is a simple closed contour in D oriented once in the counterclockwise direction. But for $n = 0, \pm 1, \pm 2, \ldots$, $\frac{f(\zeta)}{(\zeta - z_0)^{n+1}}$ is a holomorphic function of ζ on D. Since $-C_1$ and C_2 can be continuously deformed into C, it follows from Cauchy's integral theorem that

$$a_n = \frac{1}{2\pi i}\int_C \frac{f(\zeta)}{(\zeta - z_0)^{n+1}}d\zeta, \quad n = 0, \pm 1, \pm 2, \ldots,$$

and the proof is complete. □

Theorem 12.3. *Let $\sum_{n=0}^{\infty} a_n(z - z_0)^n$ and $\sum_{n=1}^{\infty} a_{-n}(z - z_0)^{-n}$ be series such that*

(1) $\sum_{n=0}^{\infty} a_n(z - z_0)^n$ converges on $\{z \in \mathbb{C} : |z - z_0| < R\}$,
(2) $\sum_{n=1}^{\infty} a_{-n}(z - z_0)^{-n}$ converges on $\{z \in \mathbb{C} : |z - z_0| > r\}$,
(3) $r < R$.

Then there exists a unique holomorphic function $w = f(z)$ on the annulus $D = \{z \in \mathbb{C} : r < |z - z_0| < R\}$ such that the Laurent series of f on D is

$$\sum_{n=0}^{\infty} a_n(z - z_0)^n + \sum_{n=1}^{\infty} a_{-n}(z - z_0)^{-n}.$$

Theorem 12.3 asserts that a series of the form

$$\sum_{n=0}^{\infty} a_n(z - z_0)^n + \sum_{n=1}^{\infty} a_{-n}(z - z_0)^{-n}$$

convergent on an annulus to a function must be the Laurent series of the function on the annulus, no matter how it is obtained.

Proof Without loss of generality, we assume that $z_0 = 0$. Let $\zeta = \frac{1}{z}$. Then $\sum_{n=1}^{\infty} a_{-n}\zeta^n$ converges on $\left\{\zeta \in \mathbb{C} : |\zeta| < \frac{1}{r}\right\}$. Let

$$H(\zeta) = \sum_{n=1}^{\infty} a_{-n}\zeta^n, \quad |\zeta| < \frac{1}{r}.$$

By Theorem 11.11, H is holomorphic on $\left\{\zeta \in \mathbb{C} : |\zeta| < \frac{1}{r}\right\}$. So, the function

$$h(z) = H\left(\frac{1}{z}\right)$$

is holomorphic on $\{z \in \mathbb{C} : |z| > r\}$ and

$$h(z) = \sum_{n=1}^{\infty} a_{-n} z^{-n}, \quad |z| > r.$$

By Theorem 11.11 again, the function

$$g(z) = \sum_{n=0}^{\infty} a_n z^n, \quad |z| < R,$$

is holomorphic. Therefore the function $w = f(z)$ given by

$$f(z) = g(z) + h(z), \quad r < |z| < R,$$

is holomorphic. It remains to prove that $\sum_{n=0}^{\infty} a_n z^n + \sum_{n=1}^{\infty} a_{-n} z^{-n}$ is the Laurent series of f on the annulus

$$D = \{z \in \mathbb{C} : r < |z| < R\}.$$

To do this, let C be a simple closed contour in D enclosing $z_0 = 0$ and oriented once in the counterclockwise direction. Then for $j = 0, \pm 1, \pm 2, \ldots$,

$$\int_C \frac{f(z)}{z^{j+1}} dz = \sum_{n=-\infty}^{\infty} a_n \int_C z^{n-j-1} dz = \begin{cases} 2\pi i \, a_j, & n = j, \\ 0, & n \neq j. \end{cases}$$

Integration term by term is permitted because the series converges uniformly on every closed subannulus of D and hence on C. Therefore

$$a_j = \frac{1}{2\pi i} \int_C \frac{f(z)}{z^{j+1}} dz, \quad j = 0, \pm 1, \pm 2, \ldots,$$

and the proof is complete. □

Example 12.4. Find the Laurent series of

$$f(z) = \frac{z^2 - 2z + 3}{z - 2}$$

on $\{z \in \mathbb{C} : |z - 1| > 1\}$.

Solution Note that $z_0 = 1$, $r = 1$ and $R = \infty$. So, f is holomorphic on the annulus $\{z \in \mathbb{C} : r < |z - 1| < R\}$. Now,

$$\frac{1}{z-2} = \frac{1}{(z-1) - 1} = \frac{1}{z-1} \frac{1}{1 - \frac{1}{z-1}}.$$

Thus, for $|z - 1| > 1$,

$$\frac{1}{z-2} = \frac{1}{z-1} \sum_{n=0}^{\infty} \frac{1}{(z-1)^n} = \sum_{n=0}^{\infty} \frac{1}{(z-1)^{n+1}}.$$

Note that

$$z^2 - 2z + 3 = z^2 - 2z + 1 + 2 = (z-1)^2 + 2$$

and hence we get

$$\begin{aligned} & f(z) \\ &= \{(z-1)^2 + 2\}\left\{\frac{1}{z-1} + \frac{1}{(z-1)^2} + \frac{1}{(z-1)^3} + \cdots\right\} \\ &= \left\{(z-1) + 1 + \frac{1}{(z-1)} + \frac{1}{(z-1)^2} + \cdots\right\} + \left\{\frac{2}{z-1} + \frac{2}{(z-1)^2} + \cdots\right\} \\ &= (z-1) + 1 + \sum_{n=1}^{\infty} \frac{3}{(z-1)^n}, \quad |z-1| > 1. \end{aligned}$$

Example 12.5. Find the Laurent series of

$$f(z) = \frac{1}{(z-1)(z-2)}$$

on $\{z \in \mathbb{C} : 1 < |z| < 2\}$ and also on $\{z \in \mathbb{C} : |z| > 2\}$.

Solution By partial fractions or inspection,

$$\frac{1}{(z-1)(z-2)} = \frac{1}{z-2} - \frac{1}{z-1}.$$

For $1 < |z| < 2$,

$$\frac{1}{z-2} = -\frac{1}{2}\frac{1}{1-\frac{z}{2}} = -\frac{1}{2}\sum_{n=0}^{\infty}\left(\frac{z}{2}\right)^n = -\sum_{n=0}^{\infty}\frac{z^n}{2^{n+1}}$$

and

$$\frac{1}{z-1} = \frac{1}{z}\frac{1}{1-\frac{1}{z}} = \frac{1}{z}\sum_{n=0}^{\infty}\frac{1}{z^n} = \sum_{n=0}^{\infty}\frac{1}{z^{n+1}}.$$

Hence for $1 < |z| < 2$,

$$f(z) = -\sum_{n=0}^{\infty}\frac{z^n}{2^{n+1}} - \sum_{n=1}^{\infty}\frac{1}{z^n}.$$

For $|z| > 2$, we still have

$$\frac{1}{z-1} = \sum_{n=0}^{\infty}\frac{1}{z^{n+1}},$$

but

$$\frac{1}{z-2} = \frac{1}{z}\frac{1}{1-\frac{2}{z}} = \frac{1}{z}\sum_{n=0}^{\infty}\left(\frac{2}{z}\right)^n = \sum_{n=0}^{\infty}\frac{2^n}{z^{n+1}}.$$

So, for $|z| > 2$,

$$f(z) = \sum_{n=1}^{\infty} \frac{2^n - 1}{z^{n+1}} = \frac{1}{z^2} + \frac{3}{z^3} + \frac{7}{z^4} + \cdots.$$

Example 12.6. Find the Laurent series of $f(z) = e^{1/z}$ at 0.

Solution Let $w = \frac{1}{z}$. Then

$$e^w = \sum_{n=0}^{\infty} \frac{w^n}{n!}$$

and hence

$$f(z) = \sum_{n=0}^{\infty} \frac{1}{n!} \frac{1}{z^n}, \quad |z| > 0.$$

Laurent series can be used to classify isolated singularities of complex-valued functions of a complex variable.

Definition 12.7. Let $w = f(z)$ be a complex-valued function. Let z_0 be a point in $\mathbb{C}$ such that

(1) f is not holomorphic at z_0,
(2) f is holomorphic on some punctured disk $\{z \in \mathbb{C} : 0 < |z - z_0| < R\}$.

Then we say that z_0 is an isolated singularity of f.

Suppose that z_0 is an isolated singularity of f. Then f is holomorphic on $\{z \in \mathbb{C} : 0 < |z - z_0| < R\}$. Suppose that

$$f(z) = \sum_{n=0}^{\infty} a_n (z - z_0)^n + \sum_{n=1}^{\infty} a_{-n} (z - z_0)^{-n}, \quad 0 < |z - z_0| < R.$$

Then we introduce the following terminology.

(1) If $a_{-n} = 0$, $n = 1, 2, \ldots$, we say that z_0 is a removable singularity.
(2) If $a_{-m} \neq 0$ for some positive integer m, but $a_{-n} = 0$ for $n > m$, we say that z_0 is a pole of order m. A pole of order 1 is called a simple pole.
(3) If $a_{-n} \neq 0$ for infinitely many positive integers n, we say that z_0 is an essential singularity.

Example 12.8. Classify the isolated singularities, if any, for each of the following functions.

(1) $w = f(z) = e^{1/z}$

(2) $w = f(z) = \frac{\sin z}{z}$
(3) $w = f(z) = \frac{e^z}{z^2}$

Solution 0 is the isolated singularity of each of the given functions. Note that

$$e^{1/z} = \sum_{n=0}^{\infty} \frac{1}{n!} \frac{1}{z^n}, \quad |z| > 0.$$

So, there are infinitely many negative powers. Therefore 0 is an essential singularity. Next,

$$\frac{\sin z}{z} = \frac{1}{z}\left\{z - \frac{z^3}{3!} + \frac{z^5}{5!} - \cdots\right\} = 1 - \frac{z^2}{3!} + \frac{z^4}{5!} - \cdots.$$

There are no negative powers. So, 0 is a removable singularity. Finally,

$$\frac{e^z}{z^2} = \frac{1}{z^2}\sum_{n=0}^{\infty} \frac{z^n}{n!} = \frac{1}{z^2} + \frac{1}{z} + \frac{1}{2!} + \frac{1}{3!}z + \frac{1}{4!}z^2 + \cdots.$$

So, 0 is a pole of order 2.

Exercises

(1) Find the Laurent series of $\frac{z+1}{z(z-4)^3}$ on $\{z \in \mathbb{C} : 0 < |z-4| < 4\}$.
(2) Find the Laurent series of $z^2 \cos\left(\frac{1}{3z}\right)$ on $\mathbb{C} - \{0\}$.
(3) Use the Laurent series of $e^{1/z}$ on the punctured plane $\mathbb{C} - \{0\}$ to compute $\frac{1}{\pi}\int_0^\pi e^{\cos\theta}\cos(\sin\theta - n\theta)\,d\theta$ for $n = 0, 1, 2, \ldots$.
(4) Let $w \in \mathbb{C}$. Prove that

$$e^{\left[\frac{w}{2}\left(z - \frac{1}{z}\right)\right]} = \sum_{n=-\infty}^{\infty} J_n(w) z^n, \quad z \in \mathbb{C} - \{0\},$$

where

$$J_n(w) = \frac{1}{2\pi}\int_0^{2\pi} e^{-in\theta} e^{iw\sin\theta}\,d\theta = \frac{1}{\pi}\int_0^\pi \cos(w\sin\theta - n\theta)\,d\theta.$$

(For $n = 0, \pm 1, \pm 2, \ldots$, the function $J_n(w)$ is called the Bessel function of the first kind and of order n. The function $e^{\left[\frac{w}{2}\left(z - \frac{1}{z}\right)\right]}$ is called the generating function of the Bessel functions of the first kind.)
(5) Compute the integral

$$\int_0^{2\pi} \cos^m\theta \cos n\theta\,d\theta$$

for all integers m and n by first computing the Laurent series of the function f on the punctured plane $\mathbb{C} - \{0\}$ given by

$$f(z) = \left(z + \frac{1}{z}\right)^m, \quad z \in \mathbb{C} - \{0\}.$$

(6) Find and classify all the isolated singularities of each of the following functions.

(a) $\frac{z^5}{z^3+z}$

(b) $z^4 \sin\left(\frac{1}{z^2}\right)$

(c) $\frac{\cos z}{z^2-1}$.

(7) Suppose that f has a pole of order m at z_0. Is z_0 an isolated singularity of f'? What kind of an isolated singularity is it?

Chapter 13

Residues

If $w = f(z)$ is a holomorphic function on and inside a simple closed contour Γ, then

$$\int_\Gamma f(z)\, dz = 0.$$

What happens if f has an isolated singularity z_0 inside Γ? It turns out that the answer is related to a number associated with the function f at z_0. The number is called the residue of f at z_0.

Definition 13.1. Let z_0 be an isolated singularity of $w = f(z)$. Then there exists a positive number R such that f is holomorphic on the punctured disk $\{z \in \mathbb{C} : 0 < |z - z_0| < R\}$. Let

$$\sum_{n=0}^{\infty} a_n (z - z_0)^n + \sum_{n=1}^{\infty} a_{-n} (z - z_0)^{-n}$$

be the Laurent series of f on $\{z \in \mathbb{C} : 0 < |z - z_0| < R\}$. Then we call the coefficient a_{-1} of $(z - z_0)^{-1}$ the residue of f at z_0 and denote it by $\operatorname{Res}(f, z_0)$.

Here are three important canonical examples.

Example 13.2. Compute the residue $\operatorname{Res}(f, z_0)$ of a function $w = f(z)$ at a removable singularity z_0.

Solution Since z_0 is a removable singularity, we get $a_{-1} = 0$. Therefore

$$\operatorname{Res}(f, z_0) = 0.$$

Example 13.3. Compute the residue $\operatorname{Res}(f, z_0)$ of a function $w = f(z)$ at a simple pole z_0.

Solution There exists a positive number R such that

$$f(z) = \sum_{n=0}^{\infty} a_n(z-z_0)^n + \sum_{n=1}^{\infty} a_{-n}(z-z_0)^{-n}, \quad 0 < |z-z_0| < R.$$

Since z_0 is a simple pole, we get

$$a_{-1} \neq 0$$

and

$$a_{-n} = 0, \quad n = 2, 3, \ldots.$$

Therefore

$$f(z) = \sum_{n=0}^{\infty} a_n(z-z_0)^n + a_{-1}(z-z_0)^{-1}, \quad 0 < |z-z_0| < R.$$

So,

$$(z-z_0)f(z) = a_{-1} + (z-z_0)\sum_{n=0}^{\infty} a_n(z-z_0)^n, \quad 0 < |z-z_0| < R.$$

Therefore

$$\lim_{z \to z_0} (z-z_0)f(z) = a_{-1}$$

and we get

$$\mathrm{Res}(f, z_0) = \lim_{z \to z_0} (z-z_0)f(z).$$

Example 13.4. Compute the residue $\mathrm{Res}(f, z_0)$ of a function f at a pole z_0 of order m.

Solution There exists a positive number R such that

$$f(z) = \sum_{n=0}^{\infty} a_n(z-z_0)^n + \sum_{n=1}^{\infty} a_{-n}(z-z_0)^{-n}, \quad 0 < |z-z_0| < R.$$

Since z_0 is a pole of order m, we get

$$a_{-m} \neq 0$$

and

$$a_{-n} = 0, \quad n > m.$$

Thus, for $0 < |z-z_0| < R$,

$$f(z) = \sum_{n=0}^{\infty} a_n(z-z_0)^n + a_{-m}(z-z_0)^{-m} + \cdots + a_{-2}(z-z_0)^{-2} + a_{-1}(z-z_0)^{-1}$$

and we get

$$
\begin{aligned}
&(z-z_0)^m f(z) \\
&= \sum_{n=0}^{\infty} a_n (z-z_0)^{n+m} + a_{-m} + \cdots + a_{-2}(z-z_0)^{m-2} + a_{-1}(z-z_0)^{m-1}.
\end{aligned}
$$

Recall the formula

$$
\frac{d^\beta}{dz^\beta} z^\alpha = \begin{cases} \binom{\alpha}{\beta} \beta! z^{\alpha-\beta}, & \beta \le \alpha, \\ 0, & \beta > \alpha. \end{cases}
$$

So, for $0 < |z - z_0| < R$,

$$
\begin{aligned}
&\frac{d^{m-1}}{dz^{m-1}} \{(z-z_0)^m f(z)\} \\
&= \sum_{n=0}^{\infty} a_n \binom{m+n}{m-1} (m-1)!(z-z_0)^{n+1} + \binom{m-1}{m-1} (m-1)! a_{-1}.
\end{aligned}
$$

Therefore

$$
\lim_{z \to z_0} \frac{d^{m-1}}{dz^{m-1}} \{(z-z_0)^m f(z)\} = (m-1)! a_{-1}
$$

and we get

$$
\operatorname{Res}(f, z_0) = \lim_{z \to z_0} \frac{1}{(m-1)!} \frac{d^{m-1}}{dz^{m-1}} \{(z-z_0)^m f(z)\}.
$$

The significance of residues is given by the following theorem, which is known as Cauchy's residue theorem.

Theorem 13.5. *Let Γ be a simple closed contour oriented once in the counterclockwise direction. Let $w = f(z)$ be a holomorphic function on and inside Γ except at the isolated singularities $z_1, z_2, \ldots, z_n$ inside Γ. Then*

$$
\int_\Gamma f(z)\, dz = 2\pi i \sum_{j=1}^{n} \operatorname{Res}(f, z_j).
$$

Proof For the sake of having simple notation, we look at the case for two isolated singularities z_1 and z_2 only. The general case is the same. Let C_1 and C_2 be non-intersecting circles inside Γ centered at z_1 and z_2 respectively and oriented once in the positive direction. Since Γ can be continuously deformed into the barbell contour as shown in Figure 13.1, Cauchy's integral theorem gives

$$
\int_\Gamma f(z)\, dz = \int_{C_1} f(z)\, dz + \int_{C_2} f(z)\, dz.
$$

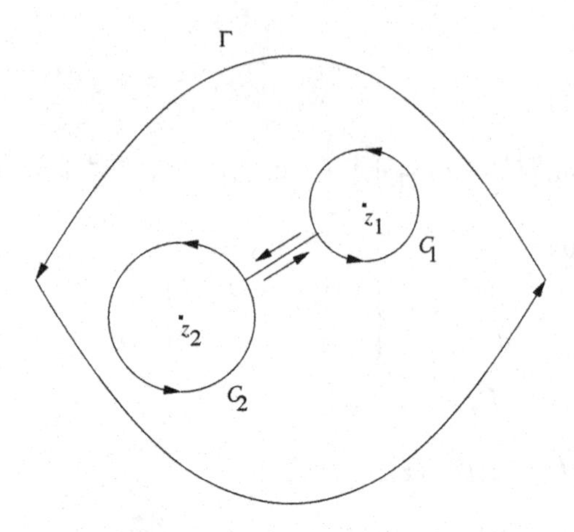

Fig. 13.1 Γ and the Barbell Contour

Now, there exists a positive number R such that

$$f(z) = \sum_{n=0}^{\infty} a_n (z - z_1)^n + \sum_{n=1}^{\infty} a_{-n} (z - z_1)^{-n}, \quad 0 < |z - z_1| < R.$$

So,

$$\begin{aligned}\int_{C_1} f(z)\, dz &= \sum_{n=0}^{\infty} a_n \int_{C_1} (z - z_1)^n dz + \sum_{n=1}^{\infty} a_{-n} \int_{C_1} (z - z_1)^{-n} dz \\ &= a_{-1} 2\pi i = 2\pi i \operatorname{Res}(f, z_1).\end{aligned}$$

Similarly,

$$\int_{C_2} f(z)\, dz = 2\pi i \operatorname{Res}(f, z_2).$$

Thus,

$$\int_{\Gamma} f(z)\, dz = 2\pi i (\operatorname{Res}(f, z_1) + \operatorname{Res}(f, z_2)).$$

□

Example 13.6. Compute $\int_C \frac{1-2z}{(z-1)(z-3)} dz$, where C is the circle with center at 0 and radius 2, and oriented once in the counterclockwise direction.

Solution Let

$$f(z) = \frac{1 - 2z}{(z - 1)(z - 3)}.$$

Then f is holomorphic everywhere except at $z = 1$ and $z = 3$. They are simple poles by inspection. Only $z = 1$ is inside C. So, by Cauchy's residue theorem,

$$\int_C \frac{1-2z}{(z-1)(z-3)} dz = 2\pi i \operatorname{Res}(f, 1).$$

But

$$\operatorname{Res}(f, 1) = \lim_{z \to 1} (z-1) f(z) = \lim_{z \to 1} \frac{1-2z}{z-3} = \frac{1}{2}.$$

Hence

$$\int_C \frac{1-2z}{(z-1)(z-3)} dz = \pi i.$$

Example 13.7. Compute $\int_C \left(ze^{1/z} + \frac{\cos z}{z^2} \right) dz$, where C is the unit circle centered at the origin and oriented once in the counterclockwise direction.

Solution Let I be the integral to be computed. Then

$$I = \int_C ze^{1/z} dz + \int_C \frac{\cos z}{z^2} dz.$$

Let $I_1 = \int_C ze^{1/z} dz$. Then

$$ze^{1/z} = z \sum_{n=0}^{\infty} \frac{1}{n!} \frac{1}{z^n} = \sum_{n=0}^{\infty} \frac{1}{n!} \frac{1}{z^{n-1}}.$$

Therefore

$$\operatorname{Res}(ze^{1/z}, 0) = \frac{1}{2}$$

and

$$I_1 = \pi i.$$

Let $I_2 = \int_C \frac{\cos z}{z^2} dz$. Then the integrand in I_2 has a pole of order 2 at the origin. So,

$$\operatorname{Res}\left(\frac{\cos z}{z^2}, 0 \right) = \lim_{z \to 0} \frac{d}{dz} (\cos z) = \lim_{z \to 0} (-\sin z) = 0.$$

Hence $I_2 = 0$ and

$$\int_C \left(ze^{1/z} + \frac{\cos z}{z^2} \right) dz = \pi i.$$

Exercises

(1) For each of the following functions, locate all the isolated singularities and compute the residue of the function at each isolated singularity.

(a) $\frac{e^z}{z(z+1)^3}$

(b) $\sin\left(\frac{1}{3z}\right)$

(c) $\frac{z^2}{1-\sqrt{z}}$ (For this exercise, $\sqrt{z}$ is the principal branch.)

(2) Use Cauchy's residue theorem to compute the following integrals.

(a) $\int_C \frac{\sin z}{z^2-4} dz$, where C is the circle with radius 5 centered at the origin and oriented once in the counterclockwise direction.

(b) $\int_C \frac{1}{z^2 \sin z} dz$, where C is the unit circle centered at the origin and oriented once in the counterclockwise direction.

(c) $\int_C e^{1/z} \sin\left(\frac{1}{z}\right) dz$, where C is the unit circle centered at the origin and oriented once in the counterclockwise direction.

Chapter 14

Trigonometric Integrals

Residues can be used to evaluate a great variety of integrals arising in many areas of mathematical analysis. The trigonometric integrals to be evaluated in this chapter are of the form

$$I = \int_0^{2\pi} U(\cos\theta, \sin\theta)\, d\theta,$$

where $U(\cos\theta, \sin\theta)$ is a rational function with real coefficients of $\cos\theta$ and $\sin\theta$. An example of such an integral is

$$\int_0^{2\pi} \frac{\sin^2\theta}{5 + 4\cos\theta} d\theta.$$

To compute I, we use the unit circle C centered at the origin and oriented once in the counterclockwise direction. We parametrize it by the equation

$$z = e^{i\theta}, \quad 0 \le \theta \le 2\pi.$$

Since

$$\cos\theta = \frac{e^{i\theta} + e^{-i\theta}}{2} = \frac{1}{2}\left(z + \frac{1}{z}\right),$$

$$\sin\theta = \frac{e^{i\theta} - e^{-i\theta}}{2i} = \frac{1}{2i}\left(z - \frac{1}{z}\right)$$

and

$$dz = ie^{i\theta} d\theta,$$

we get

$$\int_C U\left(\frac{1}{2}\left(z + \frac{1}{z}\right), \frac{1}{2i}\left(z - \frac{1}{z}\right)\right) \frac{1}{iz} dz = \int_0^{2\pi} U(\cos\theta, \sin\theta)\, d\theta.$$

Computing real integrals using residues is a powerful capability of complex analysis. Many problems in mathematical sciences require this computational skill and proficiency can be acquired only by practice. Here are some examples to serve as guidelines.

Example 14.1. Compute $I = \int_0^{2\pi} \frac{\sin^2\theta}{5+4\cos\theta}\, d\theta$.

Solution Let C be the unit circle centered at the origin and oriented once in the counterclockwise direction. Then

$$I = \int_C F(z)\, dz,$$

where

$$F(z) = \frac{\left(\frac{1}{2i}\left(z - \frac{1}{z}\right)\right)^2}{5 + 4\left(\frac{1}{2}\left(z + \frac{1}{z}\right)\right)} \frac{1}{iz} = -\frac{1}{4i} \frac{(z^2-1)^2}{z^2(2z^2+5z+2)}.$$

Therefore

$$I = -\frac{1}{4i} \int_C \frac{(z^2-1)^2}{z^2(2z^2+5z+2)} dz.$$

Let

$$f(z) = \frac{(z^2-1)^2}{z^2(2z^2+5z+2)}.$$

Then

$$f(z) = \frac{(z^2-1)^2}{2z^2\left(z+\frac{1}{2}\right)(z+2)}.$$

So, f has simple poles at $z = -\frac{1}{2}$ and $z = -2$, and a pole of order 2 at $z = 0$. Now,

$$\text{Res}\left(f, -\frac{1}{2}\right) = \lim_{z \to -\frac{1}{2}} \left(z + \frac{1}{2}\right) f(z) = \lim_{z \to -\frac{1}{2}} \frac{(z^2-1)^2}{2z^2(z+2)} = \frac{3}{4}$$

and

$$\begin{aligned} \text{Res}(f, 0) &= \lim_{z \to 0} \frac{d}{dz} \left(\frac{(z^2-1)^2}{2z^2+5z+2} \right) \\ &= \lim_{z \to 0} \frac{(2z^2+5z+2)\, 2(z^2-1)\, 2z - (z^2-1)^2(4z+5)}{(2z^2+5z+2)^2} = -\frac{5}{4}. \end{aligned}$$

Thus,

$$I = -\frac{1}{4i} 2\pi i \left(\text{Res}\left(f, -\frac{1}{2}\right) + \text{Res}(f, 0) \right) = \frac{\pi}{4}.$$

Now, we give an example of an integral that is not in the standard form, but can be transformed into one that is in the standard form.

Example 14.2. Compute $I = \int_0^\pi \frac{1}{2-\cos\theta} d\theta$.

Solution To change the integral I to one on $[0, 2\pi]$, we note that

$$I = \int_0^\pi \frac{1}{2-\cos\theta} d\theta = \int_0^\pi \frac{1}{2-\cos(2\pi-\theta)} d\theta.$$

Let $\phi = 2\pi - \theta$. Then

$$I = -\int_{2\pi}^\pi \frac{1}{2-\cos\phi} d\phi = \int_\pi^{2\pi} \frac{1}{2-\cos\phi} d\phi.$$

Therefore

$$2I = \int_0^{2\pi} \frac{1}{2-\cos\theta} d\theta.$$

So, if we let C be the unit circle centered at the origin and oriented once in the counterclockwise direction, then we get

$$2I = \int_C F(z)\, dz,$$

where

$$F(z) = \frac{1}{2 - \left(\frac{1}{2}\left(z + \frac{1}{z}\right)\right)} \frac{1}{iz} = -\frac{2}{i} \frac{1}{z^2 - 4z + 1}.$$

Let

$$f(z) = \frac{1}{z^2 - 4z + 1}.$$

Then f is holomorphic everywhere except at

$$z = \frac{4 \pm \sqrt{16-4}}{2} = \frac{4 \pm 2\sqrt{3}}{2} = 2 \pm \sqrt{3}.$$

Only the singularity $z_- = 2 - \sqrt{3}$ is inside C. Thus,

$$2I = -\frac{2}{i} \int_C f(z)\, dz = -\frac{2}{i} 2\pi i \operatorname{Res}(f, z_-).$$

But z_- is a simple pole of f. Thus,

$$\operatorname{Res}(f, z_-) = \lim_{z \to 2-\sqrt{3}} (z - z_-) f(z) = \lim_{z \to 2-\sqrt{3}} \frac{1}{z - (2+\sqrt{3})} = -\frac{1}{2\sqrt{3}}.$$

Therefore

$$2I = \frac{2\pi}{\sqrt{3}}$$

and

$$I = \frac{\pi}{\sqrt{3}}.$$

Exercises

(1) Use Cauchy's residue theorem to compute $\int_0^{2\pi} \frac{d\theta}{2+\sin\theta}$.
(2) Repeat the preceding exercise for $\int_0^{\pi} \frac{d\theta}{(3+2\cos\theta)^2}$.
(3) Let a be a complex number such that $|a| < 1$. Prove that

$$\int_0^{2\pi} \frac{1}{1-2a\cos\theta+a^2}\,d\theta = \frac{2\pi}{1-a^2}.$$

(4) What is the value of the integral in the preceding exercise when $|a| > 1$? (Hint: Let $b = \frac{1}{a}$.)
(5) Let a be a complex number such that $|a| < 1$. Prove that

$$\int_0^{2\pi} \frac{\cos\theta}{1-2a\cos\theta+a^2}\,d\theta = \frac{2\pi a}{1-a^2}.$$

Chapter 15

Cauchy Principal Values of Improper Integrals on $(-\infty, \infty)$

Let f be a continuous complex-valued function on $[0, \infty)$. Then the improper integral $\int_0^\infty f(x)\,dx$ of f on $[0, \infty)$ is defined by

$$\int_0^\infty f(x)\,dx = \lim_{b\to\infty} \int_0^b f(x)\,dx$$

if the limit exists. If f is a continuous complex-valued function on $(-\infty, 0]$, then we define the improper integral $\int_{-\infty}^0 f(x)\,dx$ of f on $(-\infty, 0]$ by

$$\int_{-\infty}^0 f(x)\,dx = \lim_{a\to-\infty} \int_a^0 f(x)\,dx$$

if the limit exists.

Let f be a continuous complex-valued function on $(-\infty, \infty)$. Then we define the improper integral $\int_{-\infty}^\infty f(x)\,dx$ of f on $(-\infty, \infty)$ by

$$\int_{-\infty}^\infty f(x)\,dx = \lim_{a\to-\infty} \int_a^0 f(x)\,dx + \lim_{b\to\infty} \int_0^b f(x)\,dx$$

provided that the two limits exist.

The following definition embodies a new concept.

Definition 15.1. Let f be a continuous complex-valued function on $(-\infty, \infty)$. Then we define the Cauchy principal value $\mathrm{pv}\int_{-\infty}^\infty f(x)\,dx$ of the improper integral of f on $(-\infty, \infty)$ by

$$\mathrm{pv}\int_{-\infty}^\infty f(x)\,dx = \lim_{\rho\to\infty} \int_{-\rho}^\rho f(x)\,dx.$$

if the limit exists.

The relationship between the improper integral and the Cauchy principal value is provided by the following theorem.

Theorem 15.2. *Let f be a continuous complex-valued function on $(-\infty, \infty)$ such that the improper integral $\int_{-\infty}^{\infty} f(x)\,dx$ exists. Then the Cauchy principal value* pv $\int_{-\infty}^{\infty} f(x)\,dx$ *of the improper integral of f on $(-\infty, \infty)$ exists and*

$$\text{pv} \int_{-\infty}^{\infty} f(x)\,dx = \int_{-\infty}^{\infty} f(x)\,dx.$$

Proof Suppose that $\int_{-\infty}^{\infty} f(x)\,dx$ exists. Then

$$\begin{aligned}\text{pv} \int_{-\infty}^{\infty} f(x)\,dx &= \lim_{\rho\to\infty} \int_{-\rho}^{\rho} f(x)\,dx \\ &= \lim_{\rho\to\infty} \int_{-\rho}^{0} f(x)\,dx + \lim_{\rho\to\infty} \int_{0}^{\rho} f(x)\,dx \\ &= \lim_{a\to-\infty} \int_{a}^{0} f(x)\,dx + \lim_{b\to\infty} \int_{0}^{b} f(x)\,dx \\ &= \int_{-\infty}^{\infty} f(x)\,dx.\end{aligned}$$

□

That the converse of the preceding theorem is false is demonstrated by the following example.

Example 15.3. Let

$$f(x) = x, \quad x \in (-\infty, \infty).$$

Compute $\int_{-\infty}^{\infty} x\,dx$ and pv $\int_{-\infty}^{\infty} x\,dx$ if they exist.

Solution Since

$$\int_{0}^{\infty} x\,dx = \lim_{b\to\infty} \int_{0}^{b} x\,dx = \lim_{b\to\infty} \frac{b^2}{2} = \infty,$$

it follows that $\int_{-\infty}^{\infty} x\,dx$ does not exist. But

$$\text{pv} \int_{-\infty}^{\infty} x\,dx = \lim_{\rho\to\infty} \int_{-\rho}^{\rho} x\,dx = 0.$$

This example demonstrates the principle that it is the cancellation effect on the symmetric interval $[-\rho, \rho]$ that makes the Cauchy principal value exist in spite of the divergence of the improper integral.

Cauchy principal values of improper integrals on $(-\infty, \infty)$ come up all the time. The following lemma will be useful to us when we come across them.

Lemma 15.4. *Let $f(z) = \frac{P(z)}{Q(z)}$, where P and Q are polynomials such that*

$$\deg(Q) - \deg(P) \geq 2.$$

Then

$$\lim_{\rho \to \infty} \int_{C_\rho^+} f(z)\, dz = 0,$$

where C_ρ^+ is the upper semicircle with radius ρ and center at the origin, oriented once in the counterclockwise direction.

Proof Write

$$f(z) = \frac{a_0 + a_1 z + a_2 z^2 + \cdots + a_m z^m}{b_0 + b_1 z + b_2 z^2 + \cdots + b_n z^n},$$

where $n - m \geq 2$. Then

$$|f(z)| = \frac{|z|^m \left| \frac{a_0}{z^m} + \frac{a_1}{z^{m-1}} + \frac{a_2}{z^{m-2}} + \cdots + a_m \right|}{|z^n| \left| \frac{b_0}{z^n} + \frac{b_1}{z^{n-1}} + \frac{b_2}{z^{n-2}} + \cdots + b_n \right|}.$$

Therefore there exists a positive number R such that

$$|f(z)| < \left(\frac{|a_m|}{|b_n|} + 1 \right) |z|^{m-n}$$

whenever $|z| \geq R$. Thus,

$$\left| \int_{C_\rho^+} f(z)\, dz \right| \leq \left(\frac{|a_m|}{|b_n|} + 1 \right) \rho^{m-n} \pi \rho = \left(\frac{|a_m|}{|b_n|} + 1 \right) \pi \rho^{m-n+1} \to 0$$

as $\rho \to \infty$. □

Example 15.5. Compute $I = \text{pv} \int_{-\infty}^{\infty} \frac{1}{x^2+1} dx$.

Solution By the definition of the Cauchy principal value, we get

$$I = \lim_{\rho \to \infty} \int_{-\rho}^{\rho} \frac{1}{x^2 + 1} dx.$$

We look at the contour integral

$$\int_{\Gamma_\rho} \frac{1}{z^2 + 1} dz,$$

where $\Gamma_\rho = C_\rho^+ + \gamma_\rho$. C_ρ^+ is the boundary of the upper semidisk with radius ρ and center at the origin, oriented once in the counterclockwise direction, and γ_ρ is the line segment from $-\rho$ to ρ. See Figure 15.1. So,

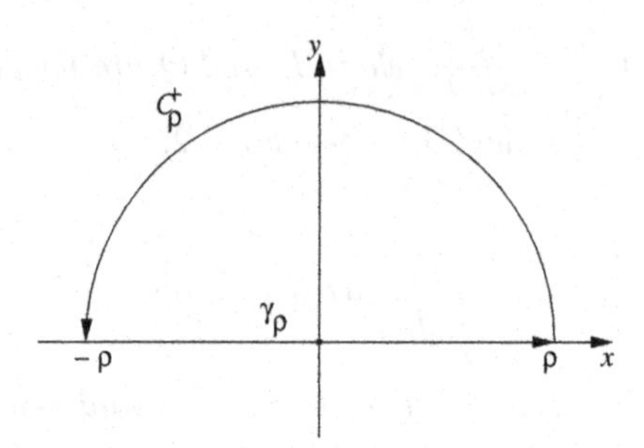

Fig. 15.1 The Contour Γ_ρ

$$\int_{\Gamma_\rho} \frac{1}{1+z^2} dz = \int_{-\rho}^{\rho} \frac{1}{x^2+1} dx + \int_{C_\rho^+} \frac{1}{z^2+1} dz.$$

Let

$$f(z) = \frac{1}{z^2+1}.$$

Then the isolated singularities are $z = i$ and $z = -i$ and they are simple poles. Thus, for all sufficiently large ρ, we get

$$\int_{\Gamma_\rho} \frac{1}{z^2+1} dz = 2\pi i \operatorname{Res}(f, i) = 2\pi i \lim_{z \to i} (z-i) f(z) = 2\pi i \lim_{z \to i} \frac{1}{z+i} = \pi.$$

Thus,

$$\pi = \int_{-\rho}^{\rho} \frac{1}{x^2+1} dx + \int_{C_\rho^+} \frac{1}{z^2+1} dz$$

for all sufficiently large ρ. Letting $\rho \to \infty$ and using Lemma 15.4, we get

$$\operatorname{pv} \int_{-\infty}^{\infty} \frac{1}{x^2+1} dx = \pi.$$

Exercises

(1) Use Cauchy's residue theorem to compute $\operatorname{pv} \int_{-\infty}^{\infty} \frac{dx}{x^2+2x+2}$.

(2) Let

$$f(x) = \frac{1}{\cosh\left(\sqrt{\frac{\pi}{2}}x\right)}, \quad x \in (-\infty, \infty).$$

Prove that

$$\frac{1}{\sqrt{2\pi}} \operatorname{pv} \int_{-\infty}^{\infty} e^{-ix\xi} f(x)\, dx = f(\xi), \quad \xi \in (-\infty, \infty).$$

(Hint: Integrate the complex-valued function $\frac{e^{-2\pi i z\xi}}{\cosh(\pi z)}$ around the rectangle with vertices at $-R$, R, $R+2i$ and $-R+2i$, which is oriented once in the counterclockwise direction. Then let $R \to \infty$ and change variables. The left hand side of the equation in this exercise is known as the Fourier transform of the function $\frac{1}{\cosh\left(\sqrt{\frac{\pi}{2}}x\right)}$. This exercise gives an example of a function whose Fourier transform is equal to itself.) Is it true that

$$\frac{1}{\sqrt{2\pi}} \int_{-\infty}^{\infty} e^{-ix\xi} f(x)\, dx = f(\xi), \quad \xi \in (-\infty, \infty)?$$

Explain your answer.

Chapter 16

Fourier Transforms of Rational Functions

Let f be a continuous complex-valued function on $(-\infty, \infty)$. Then we define the Fourier transform $\hat{f}$ of f on $(-\infty, \infty)$ by

$$\hat{f}(\xi) = \frac{1}{\sqrt{2\pi}} \text{pv} \int_{-\infty}^{\infty} e^{-ix\xi} f(x)\, dx, \quad \xi \in (-\infty, \infty),$$

provided that the Cauchy principal value exists.

Closely related to the Fourier transform $\hat{f}$ are the cosine transform and the sine transform of f given, respectively, by

$$\frac{1}{\sqrt{2\pi}} \text{pv} \int_{-\infty}^{\infty} \cos(x\xi)\, f(x)\, dx$$

and

$$\frac{1}{\sqrt{2\pi}} \text{pv} \int_{-\infty}^{\infty} \sin(x\xi)\, f(x)\, dx.$$

We give a technique in this chapter for computing Fourier transforms and their trigonometric relatives of rational functions. We begin with a lemma, which is the analog of Lemma 15.4.

Lemma 16.1. (Jordan's Lemma) *Let P and Q be polynomials such that*

$$\deg(Q) - \deg(P) \geq 1.$$

If $\xi > 0$, then

$$\lim_{\rho \to \infty} \int_{C_\rho^+} e^{iz\xi} \frac{P(z)}{Q(z)} dz = 0,$$

where C_ρ^+ is the upper semicircle with radius ρ and center at the origin, oriented once in the counterclockwise direction. If $\xi < 0$, then

$$\lim_{\rho \to \infty} \int_{C_\rho^-} e^{iz\xi} \frac{P(z)}{Q(z)} dz = 0,$$

where C_ρ^- is the lower semicircle with radius ρ and center at the origin, oriented once in the counterclockwise direction.

Proof We give a proof only for $\xi > 0$ and C_ρ^+. Let $z = \rho e^{it}$, $0 \le t \le \pi$, be the parametrization of C_ρ^+. Then

$$\int_{C_\rho^+} e^{iz\xi}\frac{P(z)}{Q(z)}dz = \int_0^\pi e^{i\xi(\rho e^{it})}\frac{P(\rho e^{it})}{Q(\rho e^{it})}\rho i e^{it}dt.$$

Let

$$g(t) = e^{i\xi(\rho e^{it})}\frac{P(\rho e^{it})}{Q(\rho e^{it})}\rho i e^{it}, \quad t \in [0, \pi].$$

Now,

$$\left|e^{i\xi(\rho e^{it})}\right| = \left|e^{i\xi\rho\cos t - \xi\rho\sin t}\right| = e^{-\xi\rho\sin t}.$$

Also, as in the proof of Lemma 15.4, we get

$$\left|\frac{P(z)}{Q(z)}\right| < \left(\frac{|a_m|}{|b_n|} + 1\right)|z|^{m-n}$$

for all sufficiently large $|z|$. Therefore

$$\left|\frac{P(\rho e^{it})}{Q(\rho e^{it})}\right| < \left(\frac{|a_m|}{|b_n|} + 1\right)\rho^{m-n}$$

for all sufficiently large ρ. Thus, for all sufficiently large ρ,

$$|g(t)| \le e^{-\xi\rho\sin t}\left(\frac{|a_m|}{|b_n|} + 1\right)\rho^{m-n+1}, \quad t \in [0, \pi].$$

Therefore

$$\left|\int_{C_\rho^+} e^{iz\xi}\frac{P(z)}{Q(z)}dz\right| = \left|\int_0^\pi g(t)\,dt\right| \le \left(\frac{|a_m|}{|b_n|} + 1\right)\rho^{m-n+1}\int_0^\pi e^{-\xi\rho\sin t}dt.$$

Note that

$$\int_0^\pi e^{-\xi\rho\sin t}dt = \int_0^{\frac{\pi}{2}} e^{-\xi\rho\sin t}dt + \int_{\frac{\pi}{2}}^\pi e^{-\xi\rho\sin t}dt.$$

Since

$$\int_{\frac{\pi}{2}}^\pi e^{-\xi\rho\sin t}dt = \int_0^{\frac{\pi}{2}} e^{-\xi\rho\sin t}dt,$$

we get

$$\int_0^\pi e^{-\xi\rho\sin t}dt = 2\int_0^{\frac{\pi}{2}} e^{-\xi\rho\sin t}dt.$$

By looking at the graphs of $y = \sin t$ and $y = \frac{2}{\pi}t$ on $\left[0, \frac{\pi}{2}\right]$ given in Figure 16.1, we get

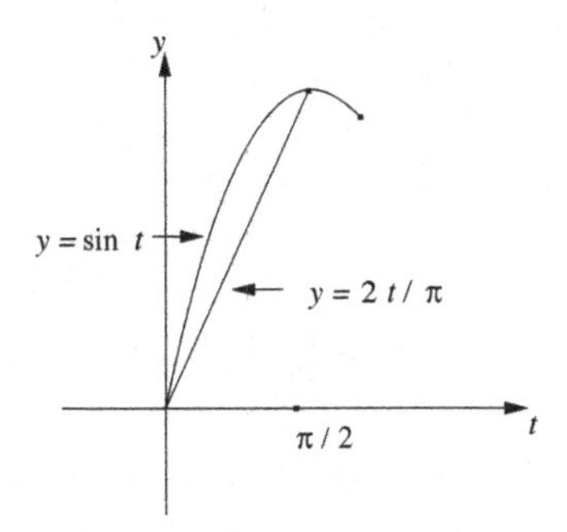

Fig. 16.1 The Sine Curve and its Secant on $\left[0, \frac{\pi}{2}\right]$

$$\sin t \geq \frac{2}{\pi} t, \quad t \in \left[0, \frac{\pi}{2}\right].$$

Hence

$$\begin{aligned}\int_0^{\pi} e^{-\xi\rho \sin t} dt &= 2 \int_0^{\frac{\pi}{2}} e^{-\xi\rho \sin t} dt \\ &\leq 2 \int_0^{\frac{\pi}{2}} e^{-\xi\rho 2t/\pi} dt \\ &= 2 \left(\frac{-\pi}{2\xi\rho}\right) (e^{-\xi\rho} - 1) < \frac{\pi}{\xi\rho}.\end{aligned}$$

So,

$$\left| \int_{C_\rho^+} e^{iz\xi} \frac{P(z)}{Q(z)} dz \right| \leq \left(\frac{|a_m|}{|b_n|} + 1 \right) \frac{\pi}{\xi} \rho^{m-n} \to 0$$

as $\rho \to \infty$. □

Example 16.2. Compute $I = \text{pv} \int_{-\infty}^{\infty} \frac{x \sin x}{1+x^2} dx$.

Solution Note that

$$I = \text{Im} \left(\text{pv} \int_{-\infty}^{\infty} \frac{xe^{ix}}{1+x^2} dx \right).$$

We look at

$$\int_{\Gamma_\rho} \frac{ze^{iz}}{1+z^2} dz,$$

where Γ_ρ is the boundary of the upper semidisk with radius ρ and center at the origin, and oriented once in the counterclockwise direction. Write $\Gamma_\rho = C_\rho^+ + \gamma_\rho$, where γ_ρ is the line segment from $-\rho$ to ρ. So,

$$\int_{\Gamma_\rho} \frac{ze^{iz}}{1+z^2} dz = \int_{C_\rho^+} \frac{ze^{iz}}{1+z^2} dz + \int_{-\rho}^{\rho} \frac{xe^{ix}}{1+x^2} dx.$$

Let

$$f(z) = \frac{ze^{iz}}{1+z^2}.$$

Then f has simple poles at $z = \pm i$. Only the pole i is inside Γ_ρ. Therefore

$$\int_{\Gamma_\rho} \frac{ze^{iz}}{1+z^2} dz = 2\pi i \operatorname{Res}(f, i) = 2\pi i \lim_{z\to i}(z-i)f(z)$$
$$= 2\pi i \lim_{z\to i} \frac{ze^{iz}}{z+i} = 2\pi i \frac{e^{-1}}{2} = \pi i e^{-1}.$$

Therefore for all sufficiently large ρ,

$$\pi i e^{-1} = \int_{C_\rho^+} \frac{ze^{iz}}{1+z^2} dz + \int_{-\rho}^{\rho} \frac{xe^{ix}}{1+x^2} dx.$$

Let $\rho \to \infty$. Then, by Jordan's lemma,

$$\text{pv} \int_{-\infty}^{\infty} \frac{xe^{ix}}{1+x^2} dx = \pi i e^{-1}.$$

Therefore

$$I = \pi e^{-1} = \frac{\pi}{e}.$$

Remark 16.3. Note that in the preceding example, we use the fact that

$$\text{pv} \int_{-\infty}^{\infty} \sin(x\xi) f(x)\, dx = \operatorname{Im} \left(\text{pv} \int_{-\infty}^{\infty} e^{ix\xi} f(x)\, dx \right).$$

This fact can be employed only if $f(x)$ is real-valued. The same remark applies to the formula

$$\text{pv} \int_{-\infty}^{\infty} \cos(x\xi) f(x)\, dx = \operatorname{Re} \left(\text{pv} \int_{-\infty}^{\infty} e^{ix\xi} f(x)\, dx \right).$$

Example 16.4. Compute $\text{pv} \int_{-\infty}^{\infty} \frac{\cos x}{x+i} dx$.

Solution In view of the preceding remark,

$$\text{pv} \int_{-\infty}^{\infty} \frac{\cos x}{x+i} dx \neq \operatorname{Re} \left(\text{pv} \int_{-\infty}^{\infty} \frac{e^{ix}}{x+i} dx \right).$$

Instead, we use

$$\cos x = \frac{e^{ix} + e^{-ix}}{2}.$$

Then

$$I = \frac{1}{2}\left(\text{pv}\int_{-\infty}^{\infty} \frac{e^{ix}}{x+i}dx\right) + \frac{1}{2}\left(\text{pv}\int_{-\infty}^{\infty} \frac{e^{-ix}}{x+i}dx\right).$$

Let

$$I_1 = \frac{1}{2}\left(\text{pv}\int_{-\infty}^{\infty} \frac{e^{ix}}{x+i}dx\right)$$

and

$$I_2 = \frac{1}{2}\left(\text{pv}\int_{-\infty}^{\infty} \frac{e^{-ix}}{x+i}dx\right).$$

For I_1, we consider

$$\int_{\Gamma_\rho^+} \frac{e^{iz}}{z+i}dz,$$

where Γ_ρ^+ is the boundary of the upper semidisk with radius ρ and center at the origin, and oriented once in the counterclockwise direction. Write $\Gamma_\rho^+ = C_\rho^+ + \gamma_\rho^+$, where γ_ρ^+ is the line segment from $-\rho$ to ρ. Let

$$f(z) = \frac{e^{iz}}{z+i}.$$

Then f is holomorphic on and inside Γ_ρ^+. So, by Cauchy's integral theorem,

$$\int_{\Gamma_\rho^+} \frac{e^{iz}}{z+i}dz = 0$$

for all positive numbers ρ. So,

$$0 = \int_{C_\rho^+} \frac{e^{iz}}{z+i}dz + \int_{-\rho}^{\rho} \frac{e^{ix}}{x+i}dx$$

for all positive numbers ρ. Let $\rho \to \infty$. Then, by Jordan's lemma, we get

$$\text{pv}\int_{-\infty}^{\infty} \frac{e^{ix}}{x+i}dx = 0$$

and hence $I_1 = 0$. For I_2, we look at

$$\int_{\Gamma_\rho^-} \frac{e^{-iz}}{z+i}dz,$$

where Γ_ρ^- is the lower semidisk with radius ρ and center at the origin, oriented once in the counterclockwise direction. Write $\Gamma_\rho^- = C_\rho^- + \gamma_\rho^-$, where γ_ρ^- is the line segment from ρ to $-\rho$. Let

$$g(z) = \frac{e^{-iz}}{z+i}.$$

Then g has a simple pole at $-i$. Thus, for all sufficiently large ρ,

$$\int_{\Gamma_\rho^-} \frac{e^{-iz}}{z+i}dz = 2\pi i \operatorname{Res}(g,-i) = 2\pi i \lim_{z\to -i}(z+i)g(z)$$
$$= 2\pi i \lim_{z\to -i} e^{-iz} = 2\pi i\, e^{-1}.$$

Therefore

$$2\pi i\, e^{-1} = \int_{C_\rho^-} \frac{e^{-iz}}{z+i}dz - \int_{-\rho}^{\rho} \frac{e^{-ix}}{x+i}dx$$

for all sufficiently large ρ. If we let $\rho \to \infty$, then, by Jordan's lemma,

$$\mathrm{pv}\int_{-\infty}^{\infty} \frac{e^{-ix}}{x+i}dx = -2\pi i e^{-1}.$$

Therefore

$$I = -\pi i e^{-1}.$$

Exercises

(1) Compute $\mathrm{pv}\int_{-\infty}^{\infty} \frac{\cos(2x)}{x-3i}dx$.

(2) Find the Fourier transform $\hat{f}$ of the function f on $(-\infty,\infty)$ given by

$$f(x) = \frac{1}{x+i}, \quad x \in (-\infty,\infty).$$

(3) Prove that for all ξ in $(-\infty,\infty)$,

$$\frac{1}{\sqrt{2\pi}}\left(\mathrm{pv}\int_{-\infty}^{\infty} e^{-ix\xi}\frac{1}{1+x^2}dx\right) = \sqrt{\frac{\pi}{2}}e^{-|\xi|}.$$

(The Fourier transform in this exercise is known as the Poisson kernel for the upper half plane and is used in the formula for the solution of the Dirichlet problem for harmonic functions on the upper half plane.)

(4) Compute the Fresnel integrals

$$\int_0^\infty \cos(x^2)\, dx$$

and

$$\int_0^\infty \sin(x^2)\, dx.$$

(Hint: Integrate e^{iz^2} around the boundary of the region S_ρ given by

$$S_\rho = \left\{ z = re^{i\theta} : 0 \le \theta \le \frac{\pi}{4},\, 0 \le r \le \rho \right\}.$$

Let $\rho \to \infty$ and use the well-known formula

$$\int_0^\infty e^{-x^2}dx = \frac{\sqrt{\pi}}{2}.$$

The Fresnel integrals are used in the formula for the solution of the Schrödinger equation for the motion of a particle in a vacuum in $\mathbb{R}^n$.)

Chapter 17

Singular Integrals on $(-\infty, \infty)$

In this chapter we look at Cauchy principal values of improper integrals $\int_{-\infty}^{\infty} f(x)\,dx$, where the integrand f has a finite number of local singularities on $(-\infty, \infty)$. These improper integrals are called singular integrals on $(-\infty, \infty)$.

Let f be a continuous complex-valued function on $(-\infty, \infty)$ except at the point c. Then the Cauchy principal value $\text{pv}\int_{-\infty}^{\infty} f(x)\,dx$ of the improper integral of f on $(-\infty, \infty)$ is given by

$$\text{pv}\int_{-\infty}^{\infty} f(x)\,dx = \lim_{\rho\to\infty, r\to 0+}\left\{\int_{-\rho}^{c-r} f(x)\,dx + \int_{c+r}^{\rho} f(x)\,dx\right\}$$

provided that the limit exists.

While Lemma 15.4 and Jordan's lemma deal with the singularity at infinity, the following lemma is designed for a local simple pole.

Lemma 17.1. *Suppose that a complex-valued function $w = f(z)$ has a simple pole at $z = c$ and T_r is the circular arc given by the parametrization*

$$z = c + re^{i\theta}, \quad \theta_1 \le \theta \le \theta_2.$$

Then

$$\lim_{r\to 0+}\int_{T_r} f(z)\,dz = i(\theta_2 - \theta_1)\text{Res}(f, c).$$

See Figure 17.1 for the contour T_r tailored for the simple pole c in Lemma 17.1.

Proof Using the Laurent series of f, there exists a positive number R such that

$$f(z) = \frac{a_{-1}}{z-c} + \sum_{n=0}^{\infty} a_n(z-c)^n, \quad 0 < |z-c| < R.$$

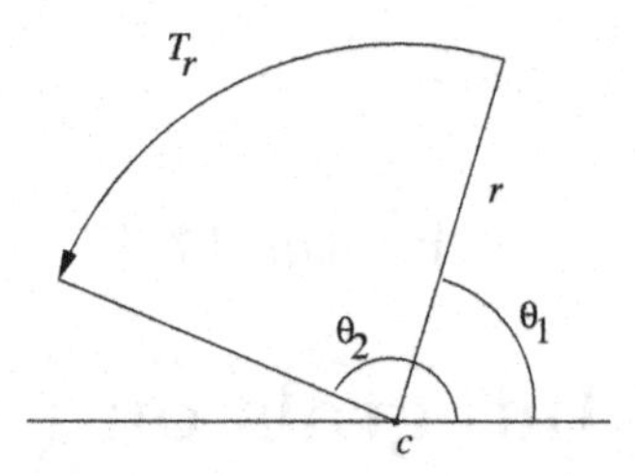

Fig. 17.1 A Circular Arc Tailored for a Simple Pole

So, for $r \in (0, R)$, we get

$$\int_{T_r} f(z)\,dz = a_{-1}\int_{T_r} \frac{1}{z-c}dz + \int_{T_r} g(z)\,dz,$$

where

$$g(z) = \sum_{n=0}^{\infty} a_n (z-c)^n.$$

Now, g is holomorphic at c. Therefore g is continuous and hence bounded on a neighborhood of c. Thus, we can suppose that there exist positive constants M and R_1 such that

$$|g(z)| \le M, \quad |z-c| < R_1.$$

Hence

$$\left|\int_{T_r} g(z)\,dz\right| \le M(\theta_2 - \theta_1)r \to 0$$

as $r \to 0+$. Also,

$$\int_{T_r} \frac{1}{z-c}dz = \int_{\theta_1}^{\theta_2} \frac{1}{re^{i\theta}} rie^{i\theta}\,d\theta = i\int_{\theta_1}^{\theta_2} d\theta = i(\theta_2 - \theta_1).$$

Therefore

$$\lim_{r\to 0+} \int_{T_r} f(z)\,dz = a_{-1} i(\theta_2 - \theta_1) = i(\theta_2 - \theta_1)\mathrm{Res}(f, c).$$

□

Example 17.2. Compute $I = \mathrm{pv}\int_{-\infty}^{\infty} \frac{e^{ix}}{x}dx$.

Solution We look at

$$\int_{\Gamma_{\rho,r}} \frac{e^{iz}}{z}dz,$$

where $\Gamma_{\rho,r}$ is the contour shown in Figure 17.2. Let $f(z) = \frac{e^{iz}}{z}$. Then f

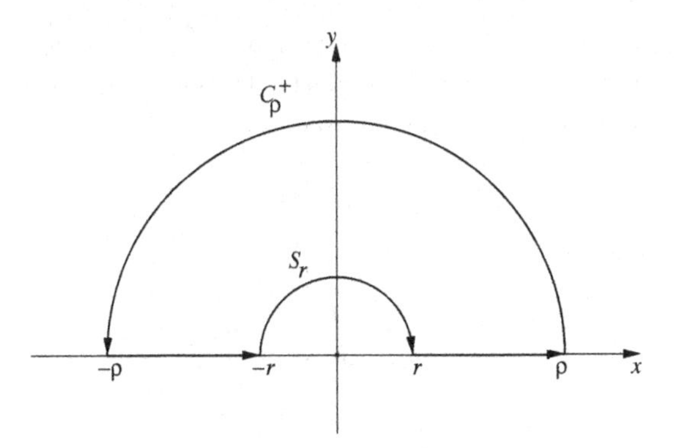

Fig. 17.2 The Contour $\Gamma_{\rho,r}$

has a simple pole at the origin, which lies outside $\Gamma_{\rho,r}$. Thus, by Cauchy's integral theorem, we get

$$\int_{\Gamma_{\rho,r}} \frac{e^{iz}}{z} dz = 0$$

for all positive numbers ρ and r with r sufficiently small and R sufficiently large. So, for $0 < r < \rho$,

$$0 = \int_{C_\rho^+} \frac{e^{iz}}{z} dz + \int_{-\rho}^{-r} \frac{e^{ix}}{x} dx + \int_{S_r} \frac{e^{iz}}{z} dz + \int_r^\rho \frac{e^{ix}}{x} dx.$$

Let $\rho \to \infty$ and $r \to 0+$. Then, by Jordan's lemma,

$$\lim_{\rho \to \infty} \int_{C_\rho^+} \frac{e^{iz}}{z} dz = 0,$$

and by Lemma 17.1,

$$\lim_{r \to 0} \int_{S_r} \frac{e^{iz}}{z} dz = -\pi i \operatorname{Res}(f, 0).$$

Hence

$$0 = \operatorname{pv} \int_{-\infty}^{\infty} \frac{e^{ix}}{x} dx = \pi i \operatorname{Res}(f, 0).$$

Therefore

$$\operatorname{pv} \int_{-\infty}^{\infty} \frac{e^{ix}}{x} dx = \pi i \operatorname{Res}(f, 0) = \pi i \lim_{z \to 0} z f(z) = \pi i \lim_{z \to 0} e^{iz} = \pi i.$$

The function $\frac{1}{x}$ in Example 17.2 is actually the kernel of the Hilbert transform. So, from Example 17.2, the Fourier transform of the kernel of the Hilbert transform at $\xi = -1$ is equal to $\sqrt{\frac{\pi}{2}}i$. A general formula for the Fourier transform of the kernel of the Hilbert transform is left as an exercise. See Exercise (2) in this chapter.

Example 17.3. Compute pv $\int_{-\infty}^{\infty} \frac{\sin x}{x} dx$.

Solution We note that

$$\mathrm{pv} \int_{-\infty}^{\infty} \frac{\sin x}{x} dx = \mathrm{Im} \left(\mathrm{pv} \int_{-\infty}^{\infty} \frac{e^{ix}}{x} dx \right) = \mathrm{Im}(\pi i) = \pi.$$

Exercises

(1) Use complex analysis to compute pv $\int_{-\infty}^{\infty} \frac{1-\cos x}{x^2} dx$.
(2) Compute

$$\frac{1}{\sqrt{2\pi}} \left(\mathrm{pv} \int_{-\infty}^{\infty} e^{-ix\xi} \frac{1}{x} dx \right), \quad \xi \in (-\infty, \infty).$$

(This is the Fourier transform of the kernel of the Hilbert transform and is usually called the symbol of the Hilbert transform.)

Chapter 18

Integrals on Branch Cuts

By way of motivation, let α be a real number, but not an integer. Let $f(z) = z^\alpha$, where the branch $z^\alpha = e^{\alpha \log_0 z}$ is taken. To recall,

$$\log_0 z = \ln|z| + i\,\theta,$$

where $0 < \theta \leq 2\pi$. Thus, f is holomorphic on the complex plane except the branch cut along the nonnegative axis. To see what may happen near the cut, let x be a positive number. Then, as $z \to x$ from the upper half plane,

$$f(z) = e^{\alpha(\ln|z| + i\,\theta)} \to e^{\alpha \ln x} = x^\alpha.$$

As $z \to x$ from the lower side,

$$f(z) = e^{\alpha(\ln|z| + i\,\theta)} \to e^{\alpha \ln x} e^{2\pi i \alpha} = x^\alpha e^{2\pi i \alpha}.$$

So, z^α behaves differently on the upper edge and the lower edge of the cut. This multi-valuedness has to be taken into consideration if we integrate along a branch cut. Examples serve best to illustrate this point.

Example 18.1. Compute $I = \int_0^\infty \frac{1}{\sqrt{x}(x+4)} dx$.

Solution We need to compute

$$\lim_{\rho \to \infty, \varepsilon \to 0+} \int_\varepsilon^\rho \frac{1}{\sqrt{x}(x+4)} dx.$$

Let

$$f(z) = \frac{1}{\sqrt{z}(z+4)},$$

where

$$\sqrt{z} = e^{\frac{1}{2}(\log_0 z)} = e^{\frac{1}{2}(\ln|z| + i\theta)}, \quad 0 < \theta < 2\pi.$$

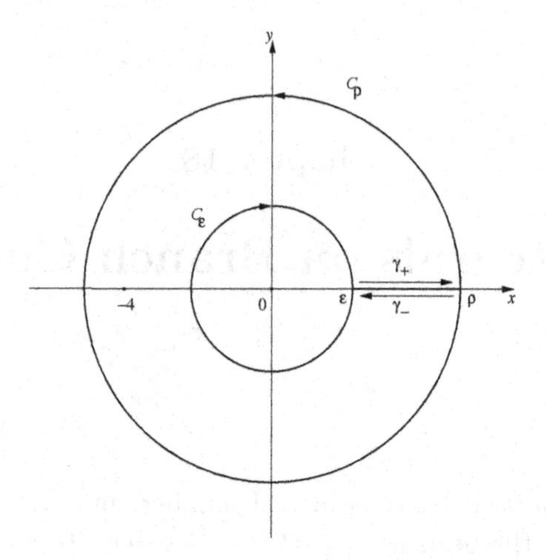

Fig. 18.1 The Contour $\Gamma_{\rho,\epsilon}$

Consider

$$\int_{\Gamma_{\rho,\varepsilon}} f(z)\,dz,$$

where $\Gamma_{\rho,\varepsilon}$ is the contour depicted in Figure 18.1. f is holomorphic on and inside the contour $\Gamma_{\rho,\varepsilon}$ except at $z=-4$, which is a simple pole. Thus,

$$\left(\int_{C_\rho}+\int_{C_\varepsilon}+\int_{\gamma_+}+\int_{\gamma_-}\right) f(z)\,dz = 2\pi i\,\mathrm{Res}(f,-4).$$

Now, on γ_+,

$$\sqrt{z}=\sqrt{x}.$$

On γ_-,

$$\sqrt{z}=\sqrt{x}e^{\pi i}=-\sqrt{x}.$$

Thus,

$$\begin{aligned}\left(\int_{\gamma_+}+\int_{\gamma_-}\right) f(z)\,dz &= \int_\varepsilon^\rho \frac{1}{\sqrt{x}(x+4)}dx - \int_\varepsilon^\rho \frac{-1}{\sqrt{x}(x+4)}dx \\ &= 2\int_\varepsilon^\rho \frac{1}{\sqrt{x}(x+4)}dx.\end{aligned}$$

So,

$$2I = \lim_{\rho\to\infty,\varepsilon\to0+}\left(\int_{\gamma_+}+\int_{\gamma_-}\right).$$

Next,

$$\left|\int_{C_\rho} f(z)\,dz\right| \le \frac{2\pi\rho}{\sqrt{\rho}(\rho-4)} \to 0$$

as $\rho \to \infty$. Also,

$$\left|\int_{C_\varepsilon} f(z)\,dz\right| \le \frac{2\pi\varepsilon}{\sqrt{\varepsilon}(4-\varepsilon)} = \frac{2\pi\sqrt{\varepsilon}}{4-\varepsilon} \to 0$$

as $\varepsilon \to 0+$. Therefore

$$2I = 2\pi i\,\mathrm{Res}(f,-4).$$

But

$$\mathrm{Res}(f,-4) = \lim_{z\to -4}\frac{1}{\sqrt{z}} = \frac{1}{e^{\frac{1}{2}(\ln|4|+i\pi)}} = e^{-\frac{1}{2}\ln 4}e^{-\frac{1}{2}i\pi} = -\frac{i}{2}.$$

Thus,

$$I = -\frac{2\pi i}{2}\frac{i}{2} = \frac{\pi}{2}.$$

Example 18.2. Compute $I = \mathrm{pv}\int_0^\infty \frac{1}{x^\alpha(x-4)}dx$, where $0<\alpha<1$.

Solution We need to compute

$$I = \lim_{\rho\to\infty,\varepsilon,\delta\to 0+}\left(\int_\varepsilon^{4-\delta} + \int_{4+\delta}^{\rho}\right)\frac{1}{x^\alpha(x-4)}dx.$$

We consider the complex integral

$$\int_{\Gamma_{\rho,\varepsilon,\delta}} \frac{1}{z^\alpha(z-4)}dz.$$

Here,

$$f(z) = z^\alpha = e^{\alpha\log_0 z} = e^{\alpha(\ln|z|+i\theta)},$$

where $0<\theta\le 2\pi$. The contour $\Gamma_{\rho,\varepsilon,\delta}$ to be used is shown in Figure 18.2. Thus, by Cauchy's integral theorem,

$$0 = \int_{\Gamma_{\rho,\varepsilon,\delta}} \frac{1}{z^\alpha(z-4)}dz = \int_{C_\rho} + \int_{C_\varepsilon} + \int_{C_{\delta,+}} + \int_{C_{\delta,-}} + \sum_{j=1}^{4}\int_{\gamma_j}.$$

On γ_1 and γ_2,

$$\frac{1}{z^\alpha(z-4)} = \frac{1}{x^\alpha(x-4)}.$$

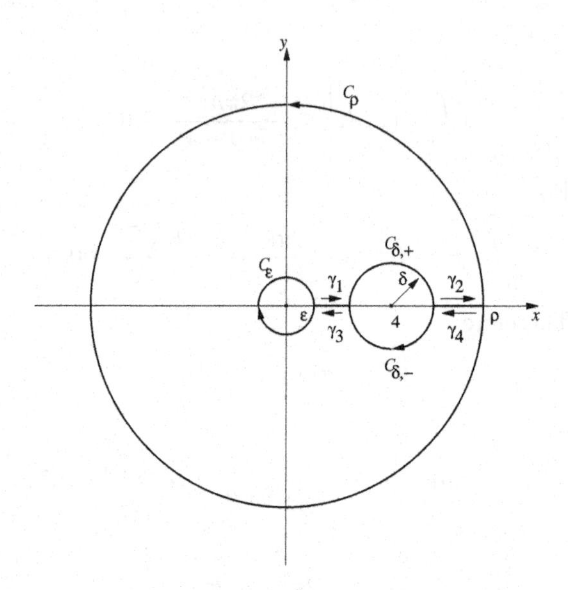

Fig. 18.2 The Contour $\Gamma_{\rho,\epsilon,\delta}$

On γ_3 and γ_4,

$$\frac{1}{z^\alpha(z-4)} = \frac{e^{-2\pi i\alpha}}{x^\alpha(x-4)}.$$

Thus,

$$\begin{aligned}\sum_{j=1}^{4}\int_{\gamma_j} &= \left(\int_\varepsilon^{4-\delta} + \int_{4+\delta}^{\rho}\right)\frac{1}{x^\alpha(x-4)}dx + \left(\int_{4-\delta}^{\varepsilon} + \int_{\rho}^{4+\delta}\right)\frac{e^{-2\pi i\alpha}}{x^\alpha(x-4)}dx \\ &= \left(\int_\varepsilon^{4-\delta} + \int_{4+\delta}^{\rho}\right)\frac{1}{x^\alpha(x-4)}dx\,(1-e^{-2\pi i\alpha}).\end{aligned}$$

As in the preceding example,

$$\lim_{\rho\to\infty}\int_{C_\rho}\frac{1}{z^\alpha(z-4)}dz = \lim_{\varepsilon\to 0+}\int_{C_\varepsilon}\frac{1}{z^\alpha(z-4)}dz = 0.$$

Letting $\rho\to\infty$ and $\varepsilon,\delta\to 0+$, we get

$$0 = \mathrm{pv}\int_0^\infty \frac{1}{x^\alpha(x-4)}dx\,(1-e^{-2\pi i\alpha}) + \lim_{\delta\to 0+}\int_{C_{\delta,+}} + \lim_{\delta\to 0+}\int_{C_{\delta,-}}.$$

Let $f(z) = \frac{1}{z^\alpha(z-4)}$. Then $z=4$ is a simple pole of f. Thus,

$$\lim_{\delta\to 0+}\int_{C_{\delta,+}} = -\pi i\,\mathrm{Res}(f,4) = -\pi i\lim_{z\to 4}\frac{1}{z^\alpha} = -\pi i\lim_{z\to 4}\frac{1}{e^{\alpha(\ln|z|+i\theta)}} = -\pi i 4^{-\alpha}.$$

Similarly,

$$\lim_{\delta\to 0+}\int_{C_{\delta,-}} = -\pi i \lim_{z\to 4}\frac{1}{z^\alpha} = \lim_{z\to 4}\frac{1}{e^{\alpha(\ln|z|+i\theta)}} = -\pi i 4^{-\alpha}e^{-2\pi i\alpha}.$$

Hence

$$\text{pv}\int_0^\infty \frac{1}{x^\alpha(x-4)}dx\,(1-e^{-2\pi i\alpha}) = \pi i 4^{-\alpha}(1+e^{-2\pi i\alpha}).$$

Therefore

$$\text{pv}\int_0^\infty \frac{1}{x^\alpha(x-4)}dx = \pi i 4^{-\alpha}\frac{1+e^{-2\pi i\alpha}}{1-e^{-2\pi i\alpha}}.$$

Exercises

(1) Compute $\int_0^\infty \frac{\sqrt{x}}{x^2+1}dx$ using a suitable contour integral. Answer: $\frac{\pi}{\sqrt{2}}$

(2) Let α be a nonzero real number in $(-1,1)$. Use a suitable contour integral to compute pv $\int_0^\infty \frac{x^\alpha}{(x+9)^2}dx$. Answer: $\frac{9^{\alpha-1}\pi\alpha}{\sin(\pi\alpha)}$

(3) Let $\alpha \in (0,1)$. Prove that

$$\int_0^\infty x^{\alpha-1}\cos x\,dx = \cos\left(\frac{\pi\alpha}{2}\right)\Gamma(\alpha),$$

where Γ is the gamma function defined by

$$\Gamma(s) = \int_0^\infty e^{-t}t^s\frac{dt}{t}, \quad s\in(0,\infty).$$

(Hint: Integrate the function $e^{-z}z^{\alpha-1}$ around the contour in Figure 18.3.)

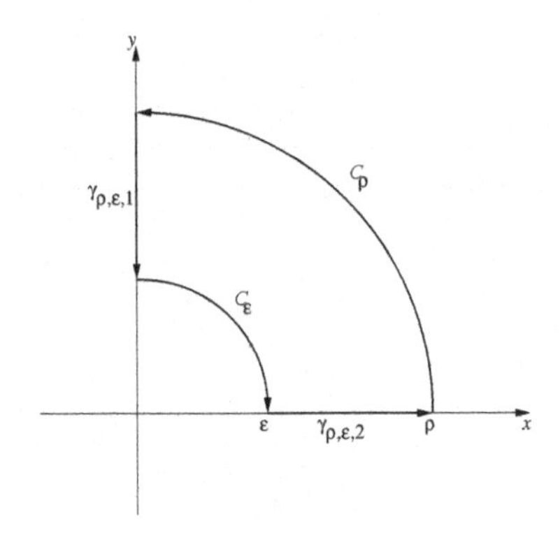

Fig. 18.3 A Suggested Contour

Chapter 19

Biholomorphisms

Let D_1 and D_2 be domains in $\mathbb{C}$. Then we say that D_1 and D_2 are biholomorphic or conformally equivalent if there exists a bijective holomorphic function $f : D_1 \to D_2$. We call a bijective holomorphic function $f : D_1 \to D_2$ a biholomorphism or a conformal mapping.

Let us recall that the two canonical domains of great interest to us are the unit disk $\mathbb{D}$ and the upper half plane $\mathbb{H}$ given by

$$\mathbb{D} = \{z \in \mathbb{C} : |z| < 1\}$$

and

$$\mathbb{H} = \{z \in \mathbb{C} : \operatorname{Im} z > 0\}.$$

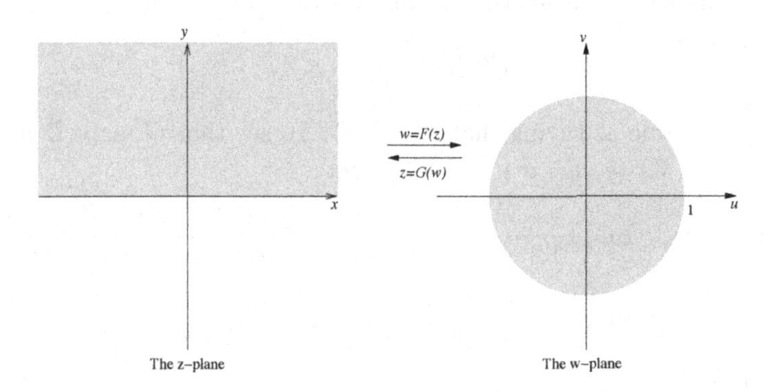

Fig. 19.1 The Cayley Transform and its Inverse

To study these domains, we make use of the functions $F : \mathbb{C} - \{-i\} \to \mathbb{C}$ and $G : \mathbb{C} - \{-1\} \to \mathbb{C}$ defined by

$$F(z) = \frac{i - z}{i + z}, \quad z \in \mathbb{C} - \{-i\},$$

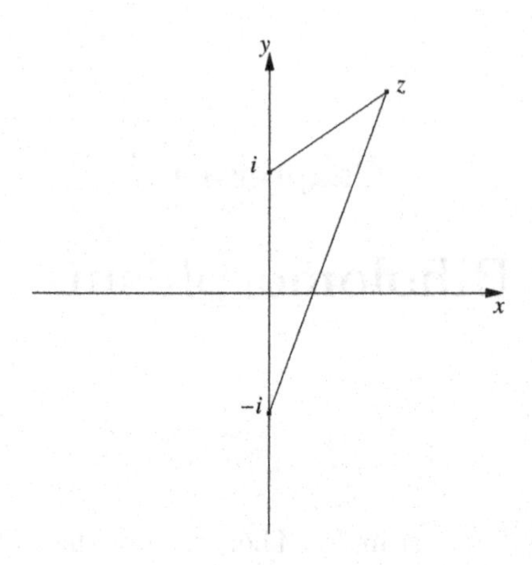

Fig. 19.2 A Geometric Argument

and

$$G(w) = i\frac{1-w}{1+w}, \quad w \in \mathbb{C} - \{-1\}.$$

Theorem 19.1. *$F : \mathbb{H} \to \mathbb{D}$ is a biholomorphism with inverse $G : \mathbb{D} \to \mathbb{H}$.*

Proof That $F : \mathbb{H} \to \mathbb{D}$ and $G : \mathbb{D} \to \mathbb{H}$ are holomorphic is obvious. Let $z \in \mathbb{H}$. Then using a geometric argument shown in Figure 19.2,

$$|F(z)| = \left|\frac{i-z}{i+z}\right| < 1,$$

which is the same as saying that $F(z) \in \mathbb{D}$. To see that G maps $\mathbb{D}$ into $\mathbb{H}$, we let $w \in \mathbb{D}$. Write $w = u + iv$ and we get

$$\begin{aligned} \operatorname{Im}(G(w)) &= \operatorname{Im}\left(i\frac{1-w}{1+w}\right) = \operatorname{Im}\left(i\frac{1-u-iv}{1+u+iv}\right) = \operatorname{Im}\frac{v+i(1-u)}{1+u+iv} \\ &= \operatorname{Im}\frac{(v+i(1-u))(1+u-iv)}{(1+u)^2+v^2} = \frac{1-u^2-v^2}{(1+u)^2+v^2} > 0. \end{aligned}$$

For all w in $\mathbb{D}$,

$$F(G(w)) = \frac{i-G(w)}{i+G(w)} = \frac{i-i\frac{1-w}{1+w}}{i+i\frac{1-w}{1+w}} = \frac{i+iw-i+iw}{i+iw+i-iw} = w.$$

Now, for all z in $\mathbb{H}$,

$$G(F(z)) = i\frac{1-F(z)}{1+F(z)} = i\frac{1-\frac{i-z}{i+z}}{1+\frac{i-z}{i+z}} = i\frac{i+z-i+z}{i+z+i-z} = z.$$

This completes the proof. □

The biholomorphism $F : \mathbb{H} \to \mathbb{D}$ given by

$$F(z) = \frac{i - z}{i + z}, \quad z \in \mathbb{H},$$

is known as the Cayley transform. Note that the Cayley transform is a special case of the so-called fractional linear transformations of the form

$$w = \frac{az + b}{cz + d}, \quad z \neq -\frac{d}{c},$$

where a, b, c and d are complex numbers such that the denominator is not a multiple of the numerator.

Remark 19.2. The Cayley transform is the biholomorphism establishing the conformal equivalence of the upper half plane $\mathbb{H}$ to the unit disk $\mathbb{D}$. So, $\mathbb{H}$ is biholomorphic to $\mathbb{D}$. This is the simplest manifestation of a fundamental fact in complex analysis, which is known as the Riemann mapping theorem and states that if D is a simply connected domain in $\mathbb{C}$ and $D \neq \mathbb{C}$, then D is biholomorphic to the unit disk $\mathbb{D}$. That the Riemann mapping theorem is not true for complex analysis on $\mathbb{C}^n$ or several complex variables is one of the remarkable features of the subject on several complex variables.

Exercises

(1) Find a biholomorphism between the disk $\{z \in \mathbb{C} : |z - i| < 1\}$ and the right half plane $\{w \in \mathbb{C} : \operatorname{Re} w > 0\}$. What is the inverse of the biholomorphism?
(2) Find a biholomorphism between the upper half plane $\mathbb{H}$ and the sector $\{w \in \mathbb{C} : 0 < \arg w < \frac{\pi}{2}\}$. What is the inverse of the biholomorphism?

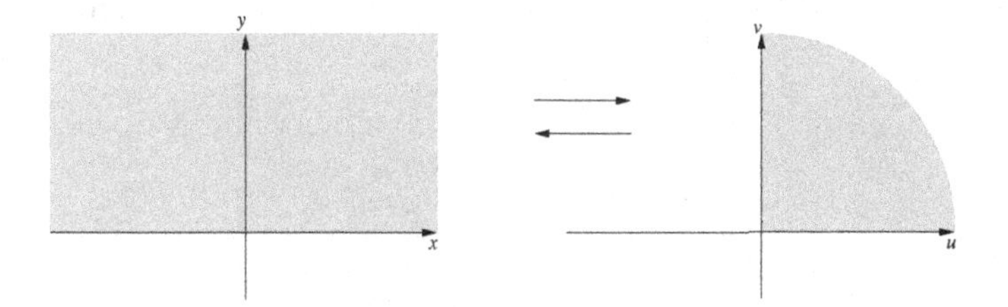

Fig. 19.3 A Biholomorphism

Chapter 20

Zeros, Maximum Modulus Principle and Schwarz's Lemma

In order to have a deeper geometric understanding of $\mathbb{D}$ and $\mathbb{H}$, we need Schwarz's lemma. To prove Schwarz's lemma, we use the maximum modulus principle, which depends on an interesting property of the zeros of holomorphic functions.

Theorem 20.1. (Unique Continuation Property) *Let Z be the zero set of a holomorphic function $w = f(z)$ on a domain D in $\mathbb{C}$. If Z has a limit point in D, then f is identically zero on D.*

The property enunciated in Theorem 20.1 is dubbed the unique continuation property in view of the fact that two holomorphic functions on a domain D agreeing on a set with a limit point in D are identically equal. In order to give a proof of this property, we first establish the following result.

Lemma 20.2. *Let z^* be a limit point of the zero set Z in the preceding theorem. Then*

$$f^{(n)}(z^*) = 0, \quad n = 0, 1, 2, \ldots.$$

In other words, f is zero on a neighborhood of z^*.

Proof Obviously, $f(z^*) = 0$. Now, by way of contradiction, suppose that k is the first positive integer for which $f^{(k)}(z^*) \neq 0$. Then there exists a positive number R such that

$$f(z) = \sum_{n=k}^{\infty} \frac{f^{(n)}(z^*)}{n!}(z - z^*)^n = (z - z^*)^k g(z), \quad |z - z^*| < R,$$

where

$$g(z) = \sum_{n=k}^{\infty} \frac{f^{(n)}(z^*)}{n!}(z - z^*)^{n-k}, \quad |z - z^*| < R.$$

Note that g is holomorphic on $\{z \in \mathbb{C} : |z - z^*| < R\}$ and $g(z^*) \neq 0$. Let $\{z_j\}_{j=1}^{\infty}$ be a sequence of points in $Z - \{z^*\}$ such that $z_j \to z^*$ as $j \to \infty$. Since

$$0 = f(z_j) = (z_j - z^*)^k g(z_j), \quad j = 1, 2, \ldots,$$

it follows that

$$g(z_j) = 0, \quad j = 1, 2, \ldots.$$

Therefore

$$g(z^*) = \lim_{j \to \infty} g(z_j) = 0.$$

This is a contradiction. □

Proof of Theorem 20.1 Let z^* be a limit point of Z in D. Then, by Lemma 20.2, $f(z) = 0$ for all z in a neighborhood of z^*. Let $\zeta \in D$. Then, by connectedness, we can join z^* and ζ by a contour Γ lying in D. Let the parametrization of Γ be

$$z = z(t), \quad a \leq t \leq b.$$

Let S be the set of all real numbers t in $[a, b]$ such that f is zero on a neighborhood of $z(t)$. Then $S \neq \phi$ because $a \in S$. S is bounded above because b is an upper bound. So, $\sup S$ exists and we denote it by μ. $\mu \in S$. Indeed, let $\{t_j\}_{j=1}^{\infty}$ be a sequence in S such that $t_j \to \mu$ as $j \to \infty$. Then for $n = 0, 1, 2, \ldots,$

$$f^{(n)}(z(t_j)) \to f^{(n)}(z(\mu))$$

as $j \to \infty$. So,

$$f^{(n)}(z(\mu)) = 0, \quad n = 0, 1, 2, \ldots.$$

This proves that f is zero on a neighborhood of $z(\mu)$ and hence $\mu \in S$. In fact, $\mu = b$. Indeed, if $\mu < b$, then there exists a positive number r such that

$$f(z) = 0, \quad |z - \mu| < r.$$

So, for some ν in $(\mu, \mu + r)$, f is zero on a neighborhood of $z(\nu)$. Thus, ν is number in S larger than μ. This contradicts the fact that μ is an upper bound of S. Therefore $b = \zeta \in S$ and $f(\zeta) = 0$. Since ζ is an arbitrary point in D, the proof is complete. □

We can now give the statement and proof of the maximum modulus principle.

Theorem 20.3. (Maximum Modulus Principle) *Let $w = f(z)$ be a nonconstant holomorphic function on a domain D in $\mathbb{C}$. Then the function $|f(z)|$ cannot attain a local maximum in D.*

Proof Suppose that $|f(z)|$ attains a local maximum at z_0 in D. Then there exists a positive number R such that

$$\{z \in \mathbb{C} : |z - z_0| < R\} \subset D$$

and

$$|f(z)| \leq |f(z_0)|, \quad |z - z_0| < R.$$

Suppose that $f(z_0) = 0$. Then

$$f(z) = 0, \quad |z - z_0| < R.$$

Hence f is identically zero on D by the unique continuation property. So, suppose that $f(z_0) \neq 0$. Let $\lambda = \frac{|f(z_0)|}{f(z_0)}$ and let C_r be the circle with radius r and center at z_0, where $0 < r < R$. Then, by the mean value property of holomorphic functions given in Exercise (3) of Chapter 10, we get

$$\begin{aligned} |f(z_0)| &= \lambda f(z_0) \\ &= \frac{1}{2\pi} \int_0^{2\pi} \lambda f(z_0 + re^{i\theta})\, d\theta \\ &= \frac{1}{2\pi} \left(\operatorname{Re} \int_0^{2\pi} \lambda f(z_0 + re^{i\theta})\, d\theta \right) \\ &= \frac{1}{2\pi} \int_0^{2\pi} \operatorname{Re}(\lambda f(z_0 + re^{i\theta}))\, d\theta. \end{aligned}$$

Therefore

$$\frac{1}{2\pi} \int_0^{2\pi} (|f(z_0)| - \operatorname{Re}(\lambda f(z_0 + re^{i\theta})))\, d\theta = 0.$$

For $0 \leq \theta \leq 2\pi$,

$$\begin{aligned} |f(z_0)| - \operatorname{Re}(\lambda f(z_0 + re^{i\theta})) &\geq |f(z_0)| - |\lambda f(z_0 + re^{i\theta})| \\ &= |f(z_0)| - |f(z_0 + re^{i\theta})| \geq 0. \end{aligned}$$

Thus,

$$|f(z_0)| = \operatorname{Re}(\lambda f(z)), \quad z \in C_r.$$

But for all z in C_r,

$$\begin{aligned} |f(z_0)|^2 &\geq |\lambda f(z)|^2 \\ &= (\operatorname{Re}(\lambda f(z)))^2 + (\operatorname{Im}(\lambda f(z)))^2 \\ &= |f(z_0)|^2 + (\operatorname{Im}(\lambda f(z)))^2. \end{aligned}$$

Therefore

$$\operatorname{Im}(\lambda f(z)) = 0, \quad z \in C_r.$$

Hence

$$\lambda f(z) = |f(z_0)|, \quad z \in C_r.$$

So, f is constant on C_r. Using the unique continuation property of holomorphic functions, we conclude that f is a constant function on D. This contradiction completes the proof of the theorem. □

The following reformulation of the maximum modulus principle is used most frequently in applications.

Theorem 20.4. *Let $w = f(z)$ be a holomorphic function on a domain D in $\mathbb{C}$. Let K be a compact subset of D. Then $\max_{z\in K} |f(z)|$ is attained at some point in the boundary of K.*

A mapping of the form $\mathbb{C} \ni z \mapsto cz \in \mathbb{C}$, where c is a complex number with $|c| = 1$, is called a rotation. So, a rotation is a mapping of the form $\mathbb{C} \ni z \mapsto e^{i\theta}z \in \mathbb{C}$, where θ is a real number. It is a rotation by θ.

We can now come to the final theme in this chapter. It expresses the fact that holomorphic functions from $\mathbb{D}$ into $\mathbb{D}$ leaving the origin fixed are rather rigid. This theme is engraved as Schwarz's lemma.

Theorem 20.5. *Let $f : \mathbb{D} \to \mathbb{D}$ be a holomorphic function with $f(0) = 0$. Then we have the following conclusions.*

(1) $|f(z)| \le |z|, \quad z \in \mathbb{D}$.
(2) If $|f(z_0)| = |z_0|$ for some z_0 in $\mathbb{D} - \{0\}$, then f is a rotation.
(3) $|f'(0)| \le 1$, and if equality holds, then f is a rotation.

Proof Since $f : \mathbb{D} \to \mathbb{D}$ is holomorphic, we use the power series expansion of f and write

$$f(z) = a_0 + a_1 z + a_2 z^2 + \cdots, \quad z \in \mathbb{D}.$$

Since $f(0) = 0$, we get $a_0 = 0$. Therefore the function $\frac{f(z)}{z}$ has a removable singularity at 0 and is holomorphic at 0. Let $r \in (0, 1)$. Then for $|z| = r$, we get

$$\left|\frac{f(z)}{z}\right| \le \frac{1}{r}.$$

So, by the maximum modulus principle, we have

$$\left|\frac{f(z)}{z}\right| \le \frac{1}{r}, \quad |z| \le r.$$

Letting $r \to 1-$, we get

$$|f(z)| \le |z|, \quad z \in \mathbb{D}.$$

This completes the proof of Part (1). For Part (2), we see that

$$\left|\frac{f(z_0)}{z_0}\right| = 1$$

and this says that the function $\frac{f(z)}{z}$ attains its maximum modulus inside $\mathbb{D}$. So, there exists a constant c such that

$$f(z) = cz, \quad z \in \mathbb{D}.$$

Thus,

$$|f(z_0)| = |c||z_0|$$

and we get $|c| = 1$. Therefore f is a rotation. For Part (3), let

$$g(z) = \frac{f(z)}{z}, \quad z \in \mathbb{D}.$$

Then $g : \mathbb{D} \to \mathbb{D}$ is a holomorphic function with

$$|g(z)| \le 1, \quad z \in \mathbb{D}.$$

Also,

$$g(0) = \lim_{z \to 0} \frac{f(z) - f(0)}{z} = f'(0).$$

If $|f'(0)| = 1$, then $|g(0)| = 1$. So, by the maximum modulus principle, there exists a constant c such that

$$g(z) = c, \quad z \in \mathbb{D}.$$

This gives $|c| = 1$ and

$$f(z) = cz, \quad z \in \mathbb{D}.$$

Therefore f is a rotation. □

Exercises

(1) Find a holomorphic function $w = f(z)$ on $\mathbb{D}$ such that

$$f\left(\frac{1}{n}\right) = \frac{1}{n^2}, \quad n = 2, 3, \ldots.$$

Is it unique? Explain your answer.

(2) Prove that there is no holomorphic function $w = f(z)$ on $\mathbb{D}$ such that

$$f\left(\frac{1}{n}\right) = \frac{(-1)^n}{n^2}, \quad n = 2, 3, \ldots.$$

(Hint: Take advantage of the preceding exercise.)

(3) Let $w = f(z)$ be a nonconstant holomorphic function on a domain D in $\mathbb{C}$. Prove that the modulus $|f|$ of f can attain a local minimum in D only at a zero of f.

(4) Use the maximum modulus principle to prove the fundamental theorem of algebra stated in Exercise (11) of Chapter 10.

(5) Let $f : \mathbb{H} \to \mathbb{C}$ be a holomorphic function such that

$$|f(z)| \leq 1, \quad z \in \mathbb{H},$$

and

$$f(i) = 0.$$

Prove that

$$|f(z)| \leq \left|\frac{i - z}{i + z}\right|, \quad z \in \mathbb{H}.$$

Chapter 21

Aut($\mathbb{D}$) and SU(1, 1)

Let D be a domain in $\mathbb{C}$. A biholomorphism $f : D \to D$ is called an automorphism of D. We denote by Aut(D) the set of all automorphisms of D. It can be shown that Aut(D) is a group with respect to the usual composition of mappings. We call Aut(D) the automorphism group of D. In the two cases of interest to us, Aut($\mathbb{D}$) and Aut($\mathbb{H}$) can be seen easily to be groups by explicit formulas in this chapter and the following chapter respectively.

Let us look at some automorphisms of $\mathbb{D}$. Let $\theta \in \mathbb{R}$. Then the rotation $r_\theta : \mathbb{C} \to \mathbb{C}$ defined by

$$r_\theta(z) = e^{i\theta} z, \quad z \in \mathbb{C},$$

is an automorphism of $\mathbb{D}$ with inverse given by $r_{-\theta}$. For another class of automorphisms of $\mathbb{D}$, we let α be a complex number with $|\alpha| < 1$ and look at the mapping $\psi_\alpha : \mathbb{D} \to \mathbb{C}$ given by

$$\psi_\alpha(z) = \frac{\alpha - z}{1 - \overline{\alpha} z}, \quad z \in \mathbb{D}.$$

We call ψ_α a Möbius transformation and we first note that it is certainly holomorphic on $\mathbb{D}$. For $|z| = 1$, we write $z = e^{i\theta}$ and we get

$$\psi_\alpha(e^{i\theta}) = \frac{\alpha - e^{i\theta}}{e^{i\theta}(e^{-i\theta} - \overline{\alpha})} = -e^{-i\theta}\frac{w}{\overline{w}},$$

where $w = \alpha - e^{i\theta}$. Therefore

$$|\psi_\alpha(z)| = 1, \quad |z| = 1.$$

By the maximum modulus principle, we get

$$|\psi_\alpha(z)| < 1, \quad z \in \mathbb{D}.$$

Now, we note that for every complex number α with $|\alpha| < 1$,

$$\begin{aligned}(\psi_\alpha \circ \psi_\alpha)(z) &= \frac{\alpha - \frac{\alpha - z}{1-\overline{\alpha}z}}{1 - \overline{\alpha}\frac{\alpha-z}{1-\overline{\alpha}z}} \\ &= \frac{\alpha - |\alpha|^2 z - \alpha + z}{1 - \overline{\alpha}z - |\alpha|^2 + \overline{\alpha}z} \\ &= \frac{(1-|\alpha|^2)z}{1-|\alpha|^2} = z\end{aligned}$$

for all z in $\mathbb{D}$. Therefore $\psi_\alpha : \mathbb{D} \to \mathbb{D}$ is its own inverse. It is also important to mention that

$$\psi_\alpha(0) = \alpha$$

and

$$\psi_\alpha(\alpha) = 0.$$

As an application of Schwarz's lemma, we can now give an explicit description of $\mathrm{Aut}(\mathbb{D})$.

Theorem 21.1. $\mathrm{Aut}(\mathbb{D}) = \left\{e^{i\theta}\psi_\alpha : \theta \in \mathbb{R}, \alpha \in \mathbb{D}\right\}.$

Proof Let $f \in \mathrm{Aut}(\mathbb{D})$. Then there exists a complex number α in $\mathbb{D}$ such that $f(\alpha) = 0$. Let $g = f \circ \psi_\alpha$. Then $g(0) = 0$. By Schwarz's lemma, we get

$$|g(z)| \leq |z|, \quad z \in \mathbb{D}.$$

Moreover,

$$g^{-1}(0) = (\psi_\alpha \circ f^{-1})(0) = \psi_\alpha(\alpha) = 0.$$

By Schwarz's lemma again,

$$|g^{-1}(w)| \leq |w|, \quad w \in \mathbb{D}.$$

This gives us

$$|z| \leq |g(z)|, \quad z \in \mathbb{D}.$$

Thus,

$$|g(z)| = |z|, \quad z \in \mathbb{D}.$$

By Schwarz's lemma again, g is a rotation and hence there exists a real number θ for which

$$g(z) = e^{i\theta}z, \quad z \in \mathbb{D}.$$

So, for all z in $\mathbb{D}$,

$$g(\psi_\alpha(z)) = e^{i\theta}\psi_\alpha(z),$$

which is the same as

$$f(\psi_\alpha(\psi_\alpha(z))) = e^{i\theta}\psi_\alpha(z).$$

Since $(\psi_\alpha \circ \psi_\alpha)(z) = z$, $z \in \mathbb{D}$, we get

$$f(z) = e^{i\theta}\psi_\alpha(z), \quad z \in \mathbb{D}.$$

□

The following corollary is immediate.

Corollary 21.2. *If $f \in \mathrm{Aut}(\mathbb{D})$ is such that $f(0) = 0$, then f is a rotation.*

Proof Let $f = e^{i\theta}\psi_\alpha$, where $\theta \in \mathbb{R}$ and $\alpha \in \mathbb{D}$. If $\alpha = 0$, then

$$f(z) = -e^{i\theta}z, \quad z \in \mathbb{D},$$

which is a rotation. Suppose $\alpha \neq 0$. Then

$$0 = f(0) = e^{i\theta}\psi_\alpha(0) = e^{i\theta}\alpha \Rightarrow \alpha = 0,$$

and this is a contradiction. □

Theorem 21.3. Aut($\mathbb{D}$) *acts transitively on $\mathbb{D}$ in the sense that for all α and β in $\mathbb{D}$, there exists an element f in* Aut($\mathbb{D}$) *such that $f(\alpha) = \beta$.*

Proof Let $f = \psi_\beta \circ \psi_\alpha$. Then $f \in \mathrm{Aut}(\mathbb{D})$ and we get

$$f(\alpha) = (\psi_\beta \circ \psi_\alpha)(\alpha) = \psi_\beta(\psi_\alpha(\alpha)) = \psi_\beta(0) = \beta.$$

□

We can give another explicit description of the elements of Aut($\mathbb{D}$) using a matrix group. To this end, we let SU(1, 1) be the set defined by

$$\mathrm{SU}(1,1) = \left\{ A = \begin{pmatrix} a & b \\ \overline{b} & \overline{a} \end{pmatrix} : a,\, b \in \mathbb{C},\, \det A = |a|^2 - |b|^2 = 1 \right\}.$$

Then SU(1, 1) is a group with respect to the usual multiplication of matrices. It is often called the pseudo-unitary group.

Let $A = \begin{pmatrix} a & b \\ \overline{b} & \overline{a} \end{pmatrix} \in \mathrm{SU}(1,1)$. Then we denote by f_A the fractional linear transformation given by

$$f_A(z) = -\frac{az + b}{\overline{b}z + \overline{a}}, \quad z \neq -\frac{\overline{a}}{\overline{b}}.$$

Here is another explicit description of Aut($\mathbb{D}$).

Theorem 21.4. $\mathrm{Aut}(\mathbb{D}) = \{f_A : A \in \mathrm{SU}(1,1)\}$.

Proof Let $f \in \mathrm{Aut}(\mathbb{D})$. Then, by Theorem 21.1, we can find a real number θ in $\mathbb{R}$ and a complex number α with $|\alpha| < 1$ such that

$$f(z) = e^{2i\theta}\frac{\alpha - z}{1 - \overline{\alpha}z}, \quad z \in \mathbb{D}.$$

Let

$$A = \begin{pmatrix} \frac{e^{i\theta}}{\sqrt{1-|\alpha|^2}} & -\frac{\alpha e^{i\theta}}{\sqrt{1-|\alpha|^2}} \\ -\frac{\overline{\alpha}e^{-i\theta}}{\sqrt{1-|\alpha|^2}} & \frac{e^{-i\theta}}{\sqrt{1-|\alpha|^2}} \end{pmatrix}.$$

Then it is clear that $A \in \mathrm{SU}(1,1)$. Moreover, for all $z \in \mathbb{D}$,

$$\begin{aligned} f_A(z) &= -\frac{\frac{e^{i\theta}}{\sqrt{1-|\alpha|^2}}z - \frac{\alpha e^{i\theta}}{\sqrt{1-|\alpha|^2}}}{-\frac{\overline{\alpha}e^{-i\theta}}{\sqrt{1-|\alpha|^2}}z + \frac{e^{-i\theta}}{\sqrt{1-|\alpha|^2}}} \\ &= -\frac{e^{i\theta}z - \alpha e^{i\theta}}{-\overline{\alpha}e^{-i\theta}z + e^{-i\theta}} \\ &= e^{2i\theta}\frac{\alpha - z}{1 - \overline{\alpha}z} = f(z). \end{aligned}$$

Conversely, let $A = \begin{pmatrix} a & b \\ \overline{b} & \overline{a} \end{pmatrix} \in \mathrm{SU}(1,1)$. We want to find a real number θ and a complex number α with $|\alpha| < 1$ such that

$$a = \frac{e^{i\theta}}{\sqrt{1-|\alpha|^2}}$$

and

$$b = -\frac{\alpha e^{i\theta}}{\sqrt{1-|\alpha|^2}}.$$

We can pick θ to be any branch of $\arg a$, *i.e.*,

$$\theta = \arg_\tau a,$$

where τ is any real number. For α, we first fix its length or modulus $|\alpha|$ by

$$|\alpha|^2 = 1 - \frac{1}{|a|^2}$$

and then α by

$$\alpha = -\sqrt{1-|\alpha|^2}\,b\,e^{-i\theta}.$$

So,

$$A = \begin{pmatrix} \frac{e^{i\theta}}{\sqrt{1-|\alpha|^2}} & -\frac{\alpha e^{i\theta}}{\sqrt{1-|\alpha|^2}} \\ -\frac{\overline{\alpha}e^{-i\theta}}{\sqrt{1-|\alpha|^2}} & \frac{e^{-i\theta}}{\sqrt{1-|\alpha|^2}} \end{pmatrix}$$

and the proof is complete. □

In spite of Theorem 21.4, the groups Aut(𝔻) and SU(1, 1) are not isomorphic. To see this, just observe that the matrices A and $-A$ in SU(1, 1) give the same element f_A in Aut(𝔻). In other words, SU(1, 1) is in some sense bigger than what we need. In fact, it is not too much bigger. To make it fit Aut(𝔻), we introduce a familiar algebraic concept.

Definition 21.5. Let G be a group. The center Z of G is defined by

$$Z = \{h \in G : gh = hg,\, g \in G\}.$$

Remark 21.6. The center of G is also known as the commutant of G in the study of operator algebras.

Theorem 21.7. *The center Z of the matrix group* SU(1, 1) *is given by*

$$Z = \left\{ \begin{pmatrix} 1 & 0 \\ 0 & 1 \end{pmatrix}, \begin{pmatrix} -1 & 0 \\ 0 & -1 \end{pmatrix} \right\}.$$

Proof Let $\begin{pmatrix} z & w \\ \overline{w} & \overline{z} \end{pmatrix} \in Z$. Then for all a and b in $\mathbb{C}$ with $|a|^2 - |b|^2 = 1$, we get

$$\begin{pmatrix} a & b \\ \overline{b} & \overline{a} \end{pmatrix} \begin{pmatrix} z & w \\ \overline{w} & \overline{z} \end{pmatrix} = \begin{pmatrix} z & w \\ \overline{w} & \overline{z} \end{pmatrix} \begin{pmatrix} a & b \\ \overline{b} & \overline{a} \end{pmatrix},$$

which is the same as

$$\begin{cases} b\overline{w} = \overline{b}w, \\ aw + b\overline{z} = bz + \overline{a}w, \\ \overline{b}z + \overline{aw} = a\overline{w} + \overline{b}\overline{z}. \end{cases}$$

From the first equation, we see that w is a real number if we let b be a nonzero real number, and hence w=0 if we put $b = i$. So, from the second or third equation, z is a real number with $z^2 = 1$. This completes the proof. □

It is clear that Z is a normal subgroup of SU(1, 1). The quotient group SU(1, 1)/Z, which is the same as $Z\backslash$SU(1, 1), is isomorphic with Aut(𝔻). The group SU(1, 1)/Z (= $Z\backslash$SU(1, 1)) is often called the projective pseudo-unitary group.

Exercises

(1) Prove that Aut(𝔻) is a group with respect to the usual composition of mappings.

(2) Find all automorphisms of $\mathbb{D}$ that leave the origin fixed. (Note that all these automorphisms form a subgroup K of Aut($\mathbb{D}$). We call K the isotropy subgroup of Aut($\mathbb{D}$).) Prove that $K\backslash\text{Aut}(\mathbb{D})$ can be identified with the unit disk $\mathbb{D}$. (The unit disk $\mathbb{D}$ so identified is known as the Poincaré disk.)

(3) Let SO(2,$\mathbb{C}$) be the subgroup of SU(1,1) given by

$$\text{SO}(2,\mathbb{C}) = \left\{ \begin{pmatrix} a & 0 \\ 0 & \overline{a} \end{pmatrix} : |a| = 1 \right\}.$$

(a) Prove that SO(2,$\mathbb{C}$) and the isotropy subgroup K of Aut($\mathbb{D}$) are isomorphic.

(b) Prove that SO(2,$\mathbb{C}$) is the isotropy subgroup of SU(1,1) in the sense that it consists of all matrices in SU(1,1) that leave the origin fixed.

(c) Prove that $\text{SO}(2,\mathbb{C})\backslash\text{SU}(1,1)$ can be identified with the unit disk $\mathbb{D}$. (The unit disk $\mathbb{D}$ so identified is also called the Poincaré disk.)

(4) Let $f : \mathbb{D} \to \mathbb{D}$ be a holomorphic function. Prove that for all z and w in $\mathbb{D}$,

$$\left| \frac{f(w) - f(z)}{1 - \overline{f(w)} f(z)} \right| \leq \left| \frac{w - z}{1 - \overline{w} z} \right|.$$

Prove that equality occurs if $f \in \text{Aut}(\mathbb{D})$. (The nonnegative number $\left|\frac{w-z}{1-\overline{w}z}\right|$ is the so-called pseudo-hyperbolic distance between the two points z and w.)

(5) Prove the Schwarz-Pick lemma, which states that if $f : \mathbb{D} \to \mathbb{D}$ is a holomorphic function, then

$$\frac{|f'(z)|}{1 - |f(z)|^2} \leq \frac{1}{1 - |z|^2}, \quad z \in \mathbb{D}.$$

(Hint: Use the preceding exercise.)

Chapter 22

Aut($\mathbb{H}$), SL($2, \mathbb{R}$) and the Iwasawa Decomposition

Let $F : \mathbb{H} \to \mathbb{D}$ be the Cayley transform studied in Chapter 19. To recall,

$$F(z) = \frac{i - z}{i + z}, \quad z \in \mathbb{H}.$$

For every φ in Aut($\mathbb{D}$), we use the Cayley transform to define the conjugation $\Gamma(\varphi)$ of φ by

$$\Gamma(\varphi) = F^{-1} \circ \varphi \circ F.$$

Then it can be proved that

$$\varphi \in \mathrm{Aut}(\mathbb{D}) \Rightarrow \Gamma(\varphi) \in \mathrm{Aut}(\mathbb{H}).$$

It can also be proved that $\Gamma : \mathrm{Aut}(\mathbb{D}) \to \mathrm{Aut}(\mathbb{H})$ is a bijection with inverse $\Gamma^{-1} : \mathrm{Aut}(\mathbb{H}) \to \mathrm{Aut}(\mathbb{D})$ given by

$$\Gamma^{-1}(\psi) = F \circ \psi \circ F^{-1}, \quad \psi \in \mathrm{Aut}(\mathbb{H}).$$

We leave the proofs of these simple assertions as exercises. In fact, we have the following result.

Theorem 22.1. $\Gamma : \mathrm{Aut}(\mathbb{D}) \to \mathrm{Aut}(\mathbb{H})$ *is a group isomorphism.*

Proof Let φ_1 and φ_2 be elements in Aut($\mathbb{D}$). Then

$$\begin{aligned}\Gamma(\varphi_1 \circ \varphi_2) &= F^{-1} \circ \varphi_1 \circ \varphi_2 \circ F \\ &= F^{-1} \circ \varphi_1 \circ F \circ F^{-1} \circ \varphi_2 \circ F \\ &= \Gamma(\varphi_1) \circ \Gamma(\varphi_2).\end{aligned}$$

This completes the proof. □

In order to give an explicit description of the elements of Aut($\mathbb{H}$), we let SL($2, \mathbb{R}$) be the set defined by

$$\mathrm{SL}(2, \mathbb{R}) = \left\{ A = \begin{pmatrix} a & b \\ c & d \end{pmatrix} : a,\ b,\ c,\ d \in \mathbb{R},\ \det A = ad - bc = 1 \right\}.$$

It can be shown easily that $\mathrm{SL}(2,\mathbb{R})$ is a group with respect to the usual multiplication of matrices. It is called the special linear group.

Let $A = \begin{pmatrix} a & b \\ c & d \end{pmatrix} \in \mathrm{SL}(2,\mathbb{R})$. Then we define the mapping $f_A : \mathbb{H} \to \mathbb{C}$ by

$$f_A(z) = \frac{az+b}{cz+d}, \quad z \in \mathbb{H}.$$

Theorem 22.2. $\mathrm{SL}(2,\mathbb{R})$ *acts transitively on* $\mathbb{H}$. *More precisely, for any two points* z *and* w *in* $\mathbb{H}$, *there exists a matrix* A *in* $\mathrm{SL}(2,\mathbb{R})$ *such that* $f_A(z) = w$.

Proof Let us first note that if $A \in \mathrm{SL}(2,\mathbb{R})$, then f_A maps $\mathbb{H}$ into $\mathbb{H}$. Indeed, let $A = \begin{pmatrix} a & b \\ c & d \end{pmatrix} \in \mathrm{SL}(2,\mathbb{R})$. Then for all z in $\mathbb{H}$,

$$\begin{aligned} \mathrm{Im}(f_A(z)) &= \mathrm{Im}\frac{az+b}{cz+d} = \mathrm{Im}\frac{(az+b)(c\bar{z}+d)}{|cz+d|^2} \\ &= \frac{(ad-bc)\mathrm{Im}\, z}{|cz+d|^2} = \frac{\mathrm{Im}\, z}{|cz+d|^2} > 0. \end{aligned} \tag{22.1}$$

It is also easy to check that if A and B are matrices in $\mathrm{SL}(2,\mathbb{R})$, then

$$f_A \circ f_B = f_{AB}.$$

This gives

$$f_A^{-1} = f_{A^{-1}}, \quad A \in \mathrm{SL}(2,\mathbb{R}).$$

Let us also note that it suffices to prove that every z in $\mathbb{H}$ can be mapped to i by f_A for some A in $\mathrm{SL}(2,\mathbb{R})$. Indeed, let $z, w \in \mathbb{H}$. Let A and B be matrices in $\mathrm{SL}(2,\mathbb{R})$ such that $f_A(z) = i$ and $f_B(w) = i$. Then

$$f_{B^{-1}A}(z) = f_{B^{-1}}(f_A(z)) = f_{B^{-1}}(i) = w.$$

We are now ready to prove that for every z in $\mathbb{H}$, there exists a matrix A in $\mathrm{SL}(2,\mathbb{R})$ such that $f_A(z) = i$. Let $d = 0$ in (22.1). Then

$$\mathrm{Im}(f_A(z)) = \frac{\mathrm{Im}\, z}{|cz|^2}.$$

Then we choose c such that $\mathrm{Im}(f_A(z)) = 1$. Let

$$A_1 = \begin{pmatrix} 0 & -c^{-1} \\ c & 0 \end{pmatrix}.$$

Then

$$\mathrm{Im}(f_{A_1}(z)) = 1.$$

Let

$$A_2 = \begin{pmatrix} 1 & b \\ 0 & 1 \end{pmatrix}.$$

Then f_{A_2} is the translation that brings $f_{A_1}(z)$ to i. Let $A = A_2 A_1$. Then

$$f_A(z) = f_{A_2 A_1}(z) = f_{A_2}(f_{A_1}(z)) = i.$$

□

The following lemma gives another way of looking at rotations.

Lemma 22.3. *Let*

$$A_\theta = \begin{pmatrix} \cos\theta & -\sin\theta \\ \sin\theta & \cos\theta \end{pmatrix}, \quad \theta \in \mathbb{R}.$$

Then

$$F \circ f_{A_\theta} \circ F^{-1} = r_{-2\theta},$$

where $F : \mathbb{H} \to \mathbb{D}$ *is the Cayley transform.*

While the proof of the lemma is straightforward, we give it here as a means to get used to the basic notions hitherto introduced.

Proof For all w in $\mathbb{D}$, we get

$$\begin{aligned}
&(F \circ f_{A_\theta} \circ F^{-1})(w) \\
&= F\left(\frac{\cos\theta F^{-1}(w) - \sin\theta}{\sin\theta F^{-1}(w) + \cos\theta}\right) \\
&= \frac{i - \frac{\cos\theta F^{-1}(w) - \sin\theta}{\sin\theta F^{-1}(w) + \cos\theta}}{i + \frac{\cos\theta F^{-1}(w) - \sin\theta}{\sin\theta F^{-1}(w) + \cos\theta}} \\
&= \frac{i\sin\theta F^{-1}(w) + i\cos\theta - \cos\theta F^{-1}(w) + \sin\theta}{i\sin\theta F^{-1}(w) + i\cos\theta + \cos\theta F^{-1}(w) - \sin\theta} \\
&= \frac{-\sin\theta\frac{1-w}{1+w} + i\cos\theta - i\cos\theta\frac{1-w}{1+w} + \sin\theta}{-\sin\theta\frac{1-w}{1+w} + i\cos\theta + i\cos\theta\frac{1-w}{1+w} - \sin\theta} \\
&= \frac{-\sin\theta + \sin\theta\, w + i\cos\theta + i\cos\theta\, w - i\cos\theta + i\cos\theta\, w + \sin\theta + \sin\theta\, w}{-\sin\theta + \sin\theta\, w + i\cos\theta + i\cos\theta\, w + i\cos\theta - i\cos\theta\, w - \sin\theta - \sin\theta\, w} \\
&= \frac{\sin\theta\, w + i\cos\theta\, w}{-\sin\theta + i\cos\theta} \\
&= (\cos^2\theta - \sin^2\theta - i\,2\sin\theta\cos\theta)w \\
&= (\cos 2\theta - i\sin 2\theta)w = r^{-2i\theta}w
\end{aligned}$$

and this completes the proof. □

Theorem 22.4. $\text{Aut}(\mathbb{H}) = \{f_A : A \in \text{SL}(2,\mathbb{R})\}$.

Proof Let $A \in \text{SL}(2,\mathbb{R})$. Then f_A is obviously holomorphic on $\mathbb{H}$. That it is bijective follows from the equation

$$f_A^{-1} = f_{A^{-1}}.$$

It remains to prove that

$$\text{Aut}(\mathbb{H}) \subseteq \{f_A : A \in \text{SL}(2,\mathbb{R})\}.$$

Let $f \in \text{Aut}(\mathbb{H})$. Let $\beta \in \mathbb{H}$ be such that $f(\beta) = i$. Let B be a matrix in $\text{SL}(2,\mathbb{R})$ such that $f_B(i) = \beta$. Let $g = f \circ f_B$. Then $g(i) = i$. Therefore $F \circ g \circ F^{-1}$ is an automorphism of $\mathbb{D}$ that leaves the origin fixed. So, $F \circ g \circ F^{-1}$ is a rotation. By Lemma 22.3, there exists a real number θ for which

$$F \circ g \circ F^{-1} = F \circ f_{A_\theta} \circ F^{-1}.$$

Therefore $g = f_{A_\theta}$ and we get

$$f = g \circ f_B^{-1} = f_{A_\theta} f_{B^{-1}} = f_{A_\theta B^{-1}}.$$

This completes the proof of the theorem. □

As in the case of $\text{SU}(1,1)$, we need to "mod" out the center of $\text{SL}(2,\mathbb{R})$ in order to get a group isomorphic with $\text{Aut}(\mathbb{H})$.

Theorem 22.5. *The center Z of* $\text{SL}(2,\mathbb{R})$ *is given by*

$$Z = \left\{ \begin{pmatrix} 1 & 0 \\ 0 & 1 \end{pmatrix}, \begin{pmatrix} -1 & 0 \\ 0 & -1 \end{pmatrix} \right\}.$$

The proof of Theorem 22.5 is similar to that of Theorem 21.7 and is hence left as an exercise. So, as in the case of $\text{Aut}(\mathbb{D})$ and $\text{SU}(1,1)$, we get

$$\text{Aut}(\mathbb{H}) = \text{SL}(2,\mathbb{R})/Z = Z\backslash\text{SL}(2,\mathbb{R}).$$

To see the upper half plane $\mathbb{H}$ in another way, we use the Iwasawa decomposition of the special linear group $\text{SL}(2,\mathbb{R})$.

Theorem 22.6. $\text{SL}(2,\mathbb{R})$ *has the Iwasawa decomposition given by*

$$\text{SL}(2,\mathbb{R}) = KAN,$$

where

$$K = \left\{ \begin{pmatrix} \cos\theta & -\sin\theta \\ \sin\theta & \cos\theta \end{pmatrix} : \theta \in \mathbb{R} \right\},$$

$$A = \left\{ \begin{pmatrix} \alpha & 0 \\ 0 & 1/\alpha \end{pmatrix} : \alpha > 0 \right\}$$

and

$$N = \left\{ \begin{pmatrix} 1 & \beta \\ 0 & 1 \end{pmatrix} : \beta \in \mathbb{R} \right\}.$$

In fact,

$$\alpha = \sqrt{a^2+c^2},\ \cos\theta = \frac{a}{\sqrt{a^2+c^2}},\ \beta = \frac{ab+cd}{a^2+c^2}.$$

Proof Let

$$g = \begin{pmatrix} a & b \\ c & d \end{pmatrix} \in \mathrm{SL}(2,\mathbb{R}).$$

Then there exists a rotation matrix

$$k = \begin{pmatrix} \cos\theta & \sin\theta \\ -\sin\theta & \cos\theta \end{pmatrix}$$

such that the column $\begin{pmatrix} a \\ c \end{pmatrix}$ is transformed into $\begin{pmatrix} \sqrt{a^2+c^2} \\ 0 \end{pmatrix}$. Thus,

$$kg = \begin{pmatrix} \sqrt{a^2+c^2} & x \\ 0 & \frac{1}{\sqrt{a^2+c^2}} \end{pmatrix},$$

where x is to be determined. From the rotation, we see that

$$\cos\theta = \frac{a}{\sqrt{a^2+c^2}},\ \sin\theta = \frac{c}{\sqrt{a^2+c^2}}.$$

Therefore

$$x = \frac{ab+cd}{\sqrt{a^2+c^2}}.$$

Thus,

$$kg = \begin{pmatrix} \sqrt{a^2+c^2} & \frac{ab+cd}{\sqrt{a^2+c^2}} \\ 0 & \frac{1}{\sqrt{a^2+c^2}} \end{pmatrix} = \begin{pmatrix} \sqrt{a^2+c^2} & 0 \\ 0 & \frac{1}{\sqrt{a^2+c^2}} \end{pmatrix} \begin{pmatrix} 1 & \frac{ab+cd}{a^2+c^2} \\ 0 & 1 \end{pmatrix}.$$

Hence

$$g = \begin{pmatrix} \cos\theta & -\sin\theta \\ \sin\theta & \cos\theta \end{pmatrix} \begin{pmatrix} \alpha & 0 \\ 0 & 1/\alpha \end{pmatrix} \begin{pmatrix} 1 & \beta \\ 0 & 1 \end{pmatrix},$$

as asserted. □

The group K is just the special orthogonal group SO(2,$\mathbb{R}$) with real entries. Thus, $\mathrm{SO}(2,\mathbb{R})\backslash\mathrm{SL}(2,\mathbb{R})$ can be identified with the group AN, which in turn can be identified with the upper half plane

$$\mathbb{H} = \{(b,a) : b \in \mathbb{R},\ a > 0\}.$$

See Exercise (6) for more details.

Exercises

(1) Prove that

$$\varphi \in \mathrm{Aut}(\mathbb{D}) \Rightarrow \Gamma(\varphi) \in \mathrm{Aut}(\mathbb{H}).$$

(2) Prove that $\Gamma : \mathrm{Aut}(\mathbb{D}) \to \mathrm{Aut}(\mathbb{H})$ is a bijection with inverse Γ^{-1} : $\mathrm{Aut}(\mathbb{H}) \to \mathrm{Aut}(\mathbb{D})$ given by

$$\Gamma^{-1}(\psi) = F \circ \psi \circ F^{-1}, \quad \psi \in \mathrm{Aut}(\mathbb{H}).$$

(3) Prove that $\mathrm{Aut}(\mathbb{H})$ is a group with respect to the usual composition of mappings.

(4) Let A and B be matrices in SL(2,$\mathbb{R}$). Prove that

$$f_A \circ f_B = f_{AB}.$$

(5) Prove that the center Z of SL(2,$\mathbb{R}$) is given by

$$Z = \left\{ \begin{pmatrix} 1 & 0 \\ 0 & 1 \end{pmatrix}, \begin{pmatrix} -1 & 0 \\ 0 & -1 \end{pmatrix} \right\}.$$

(6) Find the isotropy subgroup of all automorphisms f in $\mathrm{Aut}(\mathbb{H})$ that leave the complex number i fixed.

(7) Prove that the upper half plane $\mathbb{H}$ can be made into a group using the Iwasawa decomposition. Find the group law explicitly. (Hint: You may want to use the mapping

$$\mathbb{H} \ni (b, a) \mapsto \begin{pmatrix} \frac{1}{\sqrt{a}} & 0 \\ 0 & \sqrt{a} \end{pmatrix} \begin{pmatrix} 1 & b \\ 0 & 1 \end{pmatrix} \in AN.$$

The resulting group is often called the affine group or the $ax+b$ group, which is the underpinning of the mathematics of wavelets.)

Chapter 23

Harmonic Functions and the Schwarz Problem on $\mathbb{D}$

Let us recall that a harmonic function u on a domain D of the complex plane $\mathbb{C}$ is a real-valued function in $C^2(D)$ such that

$$\frac{\partial^2 u}{\partial x^2} + \frac{\partial^2 u}{\partial y^2} = 0$$

for all $z = (x, y)$ in D. From Exercise (9) in Chapter 10, we see that the real and imaginary parts of a holomorphic function on D are harmonic on D. This connection provides great insight into the study of harmonic functions by means of holomorphic functions. Some features of this connection are given in this chapter in the context of the unit disk $\mathbb{D}$.

Theorem 23.1. *Let u be a harmonic function on $\mathbb{D}$. Then there exists a harmonic function v on $\mathbb{D}$ such that $u + iv$ is holomorphic on $\mathbb{D}$. If w is another harmonic function on $\mathbb{D}$ for which $u + iw$ is holomorphic on $\mathbb{D}$, then $v - w$ has to be a constant function on $\mathbb{D}$.*

The function v in Theorem 23.1 is called a harmonic conjugate of u and any two harmonic conjugates of u differ by at most a constant.

Proof Let

$$f(z) = U(x, y) + iV(x, y), \quad z = (x, y) \in \mathbb{D},$$

where $U = \frac{\partial u}{\partial x}$ and $V = -\frac{\partial u}{\partial y}$. Since u is harmonic on $\mathbb{D}$, it follows that

$$\frac{\partial U}{\partial x} = \frac{\partial^2 u}{\partial x^2} = -\frac{\partial^2 u}{\partial y^2} = \frac{\partial V}{\partial y}$$

for all $z = (x, y)$ in $\mathbb{D}$. Since $u \in C^2(\mathbb{D})$, we see that

$$\frac{\partial U}{\partial y} = \frac{\partial^2 u}{\partial y \partial x} = \frac{\partial^2 u}{\partial x \partial y} = -\frac{\partial V}{\partial x}$$

for all $z = (x, y)$ in $\mathbb{D}$. So, by Theorem 5.12, f is holomorphic on $\mathbb{D}$. By Theorem 9.11,

$$\int_{\Gamma} f(z)\,dz = 0$$

for every closed contour Γ in $\mathbb{D}$. So, by Theorem 8.11, f has an antiderivative $F = \mu + i\nu$ on $\mathbb{D}$. Let $z_0 \in \mathbb{D}$. By Remark 5.9,

$$\frac{\partial \mu}{\partial x} = \frac{\partial u}{\partial x}$$

and

$$\frac{\partial \mu}{\partial y} = \frac{\partial u}{\partial y}.$$

Hence $\mu - u$ is a constant function on $\mathbb{D}$. Let this constant be C. Then the function G on $\mathbb{D}$ given by $G = F - C$ is holomorphic on $\mathbb{D}$ and

$$\operatorname{Re} G = \mu - C = u$$

on $\mathbb{D}$ and we can pick ν to be our harmonic function v. Now, let v and w be harmonic functions on $\mathbb{D}$ such that $u + iv$ and $u + iw$ are both holomorphic on $\mathbb{D}$. Then using the Cauchy–Riemann equations,

$$\frac{\partial v}{\partial x} = -\frac{\partial u}{\partial y} = \frac{\partial w}{\partial x}$$

and

$$\frac{\partial v}{\partial y} = \frac{\partial u}{\partial x} = \frac{\partial w}{\partial y}$$

for all $z = (x, y)$ in $\mathbb{D}$. So, $v - w$ has to be a constant function on $\mathbb{D}$. □

As an application of Theorem 23.1, we give the following version of Weyl's lemma. Weyl's lemma is a special case of the elliptic regularity of solutions of partial differential equations.

Theorem 23.2. (Weyl's Lemma) *Let u be a harmonic function on an open subset G of the the complex plane $\mathbb{C}$. Then $u \in C^\infty(G)$, i.e., u, v and all their partial derivatives exist and are continuous on G.*

Proof Let $z_0 \in G$. Then there exists an open disk U centered at z_0 such that $U \subset G$. By Theorem 23.1, we can find a harmonic function v on U such that the function $f = u + iv$ is holomorphic on U. So, by Theorem 10.7, all derivatives of f are holomorphic on U. By Remark 5.9, we get

$$f'(z) = \frac{\partial u}{\partial x} + i\frac{\partial v}{\partial x} = \frac{\partial v}{\partial y} - i\frac{\partial u}{\partial y}$$

and similar formulas for all derivatives of f for all $z = (x, y)$ in U. Thus, $u \in C^\infty(U)$. So, u is C^∞ on a neighborhood of z_0. Since z_0 is an arbitrary point in G, we get $u \in C^\infty(G)$, as asserted.

□

We can now give a mean value property for harmonic functions on $\mathbb{D}$.

Theorem 23.3. *Let u be a harmonic function on $\mathbb{D}$. Then for all circles $\{z \in \mathbb{C} : |z - z_0| = r\}$ lying inside $\mathbb{D}$,*

$$u(z_0) = \frac{1}{2\pi} \int_0^{2\pi} u(z_0 + r\, e^{i\theta})\, d\theta.$$

Proof Let v be a harmonic conjugate of u on $\mathbb{D}$. Then the function $f = u + iv$ is holomorphic on $\mathbb{D}$. Using the mean value property of holomorphic functions given in Exercise (3) of Chapter 10, we get

$$f(z_0) = \frac{1}{2\pi} \int_0^{2\pi} f(z_0 + r\, e^{i\theta})\, d\theta.$$

Therefore using the fact that $\operatorname{Re} f = u$, the proof is complete. □

We can also give a unique continuation property for harmonic functions on $\mathbb{D}$.

Theorem 23.4. *Let u be a harmonic function on $\mathbb{D}$ such that $u = 0$ on a nonempty and open subset U of $\mathbb{D}$. Then $u = 0$ on $\mathbb{D}$.*

Proof Let $z_0 \in U$. By Theorem 23.1, we can find a harmonic function v on $\mathbb{D}$ such that for all real numbers c, the function $u + iv + ic$ is holomorphic on $\mathbb{D}$. Let f be the holomorphic function on $\mathbb{D}$ such that

$$f(z) = u(z) + iv(z) - iv(z_0), \quad z \in \mathbb{D}.$$

Then

$$f(z_0) = u(z_0) = 0.$$

Since $\operatorname{Re} f = u = 0$ on U, it follows from the Cauchy–Riemann equations that

$$f'(z) = \frac{\partial u}{\partial x} - i\frac{\partial u}{\partial y} = 0$$

for all $z = (x, y)$ in U. Since f' is holomorphic on $\mathbb{D}$, it follows from the unique continuation property for holomorphic functions that $f' = 0$ on $\mathbb{D}$. So, f is a constant function on $\mathbb{D}$. Since $f(z_0) = 0$, we see that $f = 0$ on $\mathbb{D}$. Therefore $u = \operatorname{Re} f = 0$ on $\mathbb{D}$. □

There is also a maximum modulus principle for harmonic functions.

Theorem 23.5. *Let u be a nonconstant harmonic function on $\mathbb{D}$. Then $|u(z)|$ cannot attain a local maximum in $\mathbb{D}$.*

The proof is very similar to that of the maximum modulus principle for holomorphic functions. Indeed, suppose that $|u(z)|$ attains its maximum at z_0 in $\mathbb{D}$. Then there exists a positive number R such that

$$\{z \in \mathbb{C} : |z - z_0| < R\} \subset \mathbb{D}$$

and

$$|u(z)| \leq |u(z_0)|, \quad |z - z_0| < R.$$

Suppose that $u(z_0) = 0$. Then

$$u(z) = 0, \quad |z - z_0| < R.$$

So, by the unique continuation property for harmonic functions on $\mathbb{D}$ given by Theorem 23.4, u is identically zero on $\mathbb{D}$. Now, suppose that $u(z_0) \neq 0$. Let $\lambda = \frac{|u(z_0)|}{u(z_0)}$ and let $r \in (0, R)$. Then the mean value property for harmonic functions on $\mathbb{D}$ given by Theorem 23.3 gives

$$|u(z_0)| = \lambda u(z_0) = \frac{1}{2\pi} \int_0^{2\pi} \lambda u(z_0 + re^{i\theta})\, d\theta.$$

Therefore

$$\frac{1}{2\pi} \int_0^{2\pi} (|u(z_0)| - \lambda u(z_0 + re^{i\theta}))\, d\theta = 0.$$

For $0 \leq \theta \leq 2\pi$,

$$|u(z_0)| - \lambda u(z_0 + re^{i\theta}) \geq |u(z_0)| - |\lambda u(z_0 + re^{i\theta})| = |u(z_0)| - |u(z_0 + re^{i\theta})| \geq 0.$$

Thus,

$$|u(z_0)| = \lambda u(z), \quad |z - z_0| = r.$$

Since $r \in (0, R)$ is arbitrary, we get

$$u(z) = \frac{|u(z_0)|}{\lambda}, \quad 0 < |z - z_0| < R.$$

So, using the unique continuation property for harmonic functions on $\mathbb{D}$ given by Theorem 23.4, u is equal to a constant function on $\mathbb{D}$. This contradiction completes the proof.

Corollary 23.6. *Let u be a harmonic function on $\mathbb{D}$. Let K be a compact subset of $\mathbb{D}$. Then $\max_{z \in K} |u(z)|$ is attained at some point in the boundary of K.*

We can now give a glimpse of how complex analysis can shed light on some interesting problems for partial differential equations at least in the setting of $\mathbb{R}^2$. In keeping the prerequisites minimal for a thorough study of the book, we look at the Schwarz problem for $\mathbb{D}$ only.

Let f be a continuous real-valued function on the boundary $\partial\mathbb{D}$ of $\mathbb{D}$. The Schwarz problem is the problem of finding all holomorphic functions u on $\mathbb{D}$ such that

$$\operatorname{Re} u(re^{i\theta}) \to f(e^{i\theta})$$

uniformly with respect to θ in $[0, 2\pi]$ as $r \to 1-$.

In order to get some idea of what a solution looks like, let u be a holomorphic function on $\overline{\mathbb{D}}$ such that

$$\operatorname{Re} u = f$$

on $\partial\mathbb{D}$. Then

$$2f(w) = u(w) + \overline{u(w)}, \quad w \in \partial\mathbb{D}.$$

Therefore

$$2f(w) = \sum_{m=0}^{\infty} a_m w^m + \sum_{m=0}^{\infty} \overline{a_m w^m}, \quad w \in \partial\mathbb{D},$$

and the convergence of each series is uniform on $\partial\mathbb{D}$. Therefore for each nonnegative integer n,

$$2\int_{\partial\mathbb{D}} \frac{f(w)}{w^n}\frac{dw}{w} = \sum_{m=0}^{\infty} a_m \int_{\partial\mathbb{D}} \frac{w^m}{w^n}\frac{dw}{w} + \sum_{m=0}^{\infty} \overline{a_m} \int_{\partial\mathbb{D}} \frac{\overline{w^m}}{w^n}\frac{dw}{w}. \tag{23.1}$$

But

$$\int_{\partial\mathbb{D}} \frac{w^m}{w^n}\frac{dw}{w} = \int_0^{2\pi} e^{i(m-n)\theta} i\, d\theta = \begin{cases} 2\pi i, & m = n, \\ 0, & m \neq n. \end{cases} \tag{23.2}$$

Moreover,

$$\int_{\partial\mathbb{D}} \frac{\overline{w^m}}{w^n}\frac{dw}{w} dw = \int_0^{2\pi} e^{-i(m+n)\theta} i\, d\theta = \begin{cases} 2\pi i, & m = n = 0, \\ 0, & \text{otherwise}. \end{cases} \tag{23.3}$$

So, putting (23.2) and (23.3) into (23.1), we get

$$\frac{1}{\pi i}\int_{\partial\mathbb{D}} \frac{f(w)}{w^n}\frac{dw}{w} = \begin{cases} a_0 + \overline{a_0}, & n = 0, \\ a_n, & n \geq 1. \end{cases} \tag{23.4}$$

Therefore for all z in $\mathbb{D}$,

$$u(z) = \frac{1}{\pi i}\sum_{n=0}^{\infty} \int_{\partial\mathbb{D}} f(w)\left(\frac{z}{w}\right)^n \frac{dw}{w} - \overline{a_0}.$$

But for all z in $\mathbb{D}$ and all w in $\partial\mathbb{D}$,

$$\sum_{n=0}^{\infty}\left(\frac{z}{w}\right)^n = \frac{1}{1-\frac{z}{w}} = \frac{w}{w-z}.$$

This gives

$$u(z) = \frac{1}{\pi i}\int_{\partial\mathbb{D}} \frac{f(w)}{w-z}dw - \overline{a_0}, \quad z \in \mathbb{D}.$$

Now, by (23.4), $\operatorname{Re} a_0 = \frac{1}{2\pi i}\int_{\partial\mathbb{D}} \frac{f(w)}{w}dw$ and therefore for all z in $\mathbb{D}$,

$$\begin{aligned} u(z) &= \frac{1}{\pi i}\int_{\partial\mathbb{D}} \frac{f(w)}{w-z}dw - \operatorname{Re} a_0 + i\operatorname{Im} a_0 \\ &= \frac{1}{\pi i}\int_{\partial\mathbb{D}} \frac{f(w)}{w-z}dw - \frac{1}{2\pi i}\int_{\partial\mathbb{D}} \frac{f(w)}{w}dw + i\operatorname{Im} a_0 \\ &= \frac{1}{2\pi i}\int_{\partial\mathbb{D}} f(w)\frac{w+z}{w-z}\frac{dw}{w} + i\operatorname{Im} a_0. \end{aligned}$$

We can now "solve" the Schwarz problem.

Theorem 23.7. *Let f be a continuous real-valued function on $\partial\mathbb{D}$. Then the function u given by*

$$u(z) = \frac{1}{2\pi i}\int_{\partial\mathbb{D}} f(w)\frac{w+z}{w-z}\frac{dw}{w} + ic, \quad z \in \mathbb{D}, \tag{23.5}$$

where c is an arbitrary real number, is a solution of the Schwarz problem.

Proof We need to prove that u is holomorphic on $\mathbb{D}$ and

$$u(re^{i\theta}) \to f(e^{i\theta})$$

uniformly with respect to all θ in $[0, 2\pi]$ as $r \to 1-$. Let $z \in \mathbb{D}$. Then for all w in $\partial\mathbb{D}$,

$$\begin{aligned} \frac{w+z}{w-z} &= \frac{w}{w-z} + \frac{z}{w-z} = \frac{1}{1-\frac{z}{w}} + \frac{z}{w}\frac{1}{1-\frac{z}{w}} \\ &= \sum_{n=0}^{\infty}\left(\frac{z}{w}\right)^n + \sum_{n=1}^{\infty}\left(\frac{z}{w}\right)^n = 1 + 2\sum_{n=1}^{\infty}\left(\frac{z}{w}\right)^n, \end{aligned}$$

where the convergence of the series is uniform with respect to w in $\partial\mathbb{D}$. So,

$$\begin{aligned} u(z) &= \frac{1}{2\pi i}\int_{\partial\mathbb{D}} f(w)\frac{w+z}{w-z}\frac{dw}{w} \\ &= \frac{1}{2\pi i}\int_{\partial\mathbb{D}} f(w)\frac{dw}{w} + \frac{1}{\pi i}\left(\sum_{n=1}^{\infty}\int_{\partial\mathbb{D}} \frac{f(w)}{w^n}\frac{dw}{w}\right) z^n \end{aligned}$$

for all z in $\mathbb{D}$. Thus, u is holomorphic on $\mathbb{D}$. To prove that

$$\operatorname{Re} u(re^{i\theta}) \to f(e^{i\theta})$$

uniformly with respect to θ in $[0, 2\pi]$ as $r \to 1-$, we use polar coordinates and write

$$z = re^{i\theta}, \quad 0 < r < 1, 0 \le \theta \le 2\pi,$$

and

$$w = e^{i\phi}, \quad 0 \le \phi \le 2\pi.$$

Then, letting $F(\theta) = f(e^{i\theta})$, we get

$$u(re^{i\theta}) = \frac{1}{2\pi} \int_0^{2\pi} F(\phi) \frac{e^{i\phi} + re^{i\theta}}{e^{i\phi} - re^{i\theta}} d\phi.$$

So,

$$\operatorname{Re} u(re^{i\theta}) = \frac{1}{2\pi} \int_0^{2\pi} F(\phi) \frac{1 - r^2}{1 - 2r\cos(\theta - \phi) + r^2} d\phi.$$

Since F is uniformly continuous on $[0, 2\pi]$, it follows that for every positive number ε, there exists a positive number δ such that

$$|F(\phi) - F(\theta)| < \frac{\varepsilon}{3}$$

for all ϕ and θ in $[0, 2\pi]$ with $|\phi - \theta| < \delta$. By Exercise (3) in Chapter 14, we have

$$\begin{aligned}
&|\operatorname{Re} u(re^{i\theta}) - F(\theta)| \\
&= \left| \frac{1}{2\pi} \int_0^{2\pi} (F(\phi) - F(\theta)) \frac{1 - r^2}{1 - 2r\cos(\theta - \phi) + r^2} d\phi \right| \\
&\le \frac{1}{2\pi} \int_0^{2\pi} |F(\phi) - F(\theta)| \frac{1 - r^2}{1 - 2r\cos(\theta - \phi) + r^2} d\phi \\
&= \frac{1}{2\pi} \left(\int_{|\phi-\theta|<\delta} + \int_{\delta \le |\phi-\theta| \le 2\pi-\delta} + \int_{2\pi-\delta<|\phi-\theta|\le 2\pi} \right) \frac{|F(\phi) - F(\theta)|(1 - r^2)}{1 - 2r\cos(\theta - \phi) + r^2} d\phi. \\
&< \frac{2\varepsilon}{3} + \frac{1}{2\pi} \int_{\delta \le |\phi-\theta| \le 2\pi-\delta} |F(\phi) - F(\theta)| \frac{1 - r^2}{1 - 2r\cos(\varphi - \phi) + r^2} d\phi. \qquad (23.6)
\end{aligned}$$

In order to estimate $\frac{1}{2\pi} \int_{\delta \le |\phi-\theta| \le 2\pi-\delta} |F(\phi) - F(\theta)| \frac{1-r^2}{1-2r\cos(\theta-\phi)+r^2} d\phi$, we first note that

$$\frac{1}{1 - 2r\cos(\theta - \phi) + r^2} \le \frac{1}{1 - 2r\cos\delta + r^2}.$$

Since $1 - 2r\cos\delta + r^2$ has the absolute minimum at $r = \cos\delta$, it follows that

$$\frac{1}{1 - 2r\cos(\theta - \phi) + r^2} \leq \frac{1}{1 - \cos^2\delta}.$$

Thus, letting $M = \sup_{0\leq\phi\leq 2\pi} |F(\phi)|$, we obtain

$$\frac{1}{2\pi}\int_{\delta\leq|\phi-\theta|\leq 2\pi-\delta} |F(\phi) - F(\varphi)| \frac{1-r^2}{1 - 2r\cos(\theta - \phi) + r^2} d\phi$$
$$\leq \frac{M}{\pi}\int_{\delta\leq|\phi-\theta|\leq 2\pi-\delta} \frac{1-r^2}{1-\cos^2\delta} d\phi \leq 2M\frac{1-r^2}{1-\cos^2\delta}.$$

So, there exists a positive number δ_0 such that

$$\frac{1}{2\pi}\int_{\delta\leq|\phi-\theta|\leq 2\pi-\delta} |F(\phi) - F(\theta)| \frac{1-r^2}{1 - 2r\cos(\phi - \theta) + r^2} d\phi < \frac{\varepsilon}{3}$$

whenever $0 < |r - 1| < \delta_0$. So, by (23.6),

$$0 < |r - 1| < \delta_0 \Rightarrow |\mathrm{Re}\, u(re^{i\theta}) - f(e^{i\theta})| < \frac{2\varepsilon}{3} + \frac{\varepsilon}{3} = \varepsilon.$$

□

Theorem 23.8. *Every solution u of the Schwarz problem is of the form (23.5).*

Proof Let u and v be two solutions of the Schwarz problem. Let $U = \mathrm{Re}\, u$ and $V = \mathrm{Re}\, v$. Let $W = U - V$. Then W is a harmonic function on $\mathbb{D}$ such that

$$W(re^{i\theta}) = U(re^{i\theta}) - V(re^{i\theta}) \to f(e^{i\theta}) - f(e^{i\theta}) = 0$$

uniformly with respect to θ on $[0, 2\pi]$ as $r \to 1-$. Suppose that, by way of contradiction, there exists a point z_0 in $\mathbb{D}$ such that $W(z_0) \neq 0$. Then there exists a number ρ in $(0, 1)$ such that $|z_0| < \rho$ and

$$|W(re^{i\theta})| < \frac{|W(z_0)|}{2}$$

whenever $\rho \leq r < 1$. Now, by the maximum modulus principle for harmonic functions given in Corollary 23.6, $\max_{|z|\leq\rho} |W(z)|$ has to be attained at some point in the circle $\{z \in \mathbb{C} : |z| = \rho\}$ and this is a contradiction.

□

Remark 23.9. For $0 < r < 1$, the function P_r defined on $[0, 2\pi]$ by

$$P_r(\theta) = \frac{1}{2\pi}\frac{1-r^2}{1 - 2r\cos\theta + r^2}, \quad 0 \leq \theta \leq 2\pi,$$

is known as the Poisson kernel for the unit disk $\mathbb{D}$.

Remark 23.10. For $0 < r < 1$, the function $P_r * f$ on $[0, 2\pi]$ defined by

$$(P_r * f)(\theta) = \frac{1}{2\pi}\int_0^{2\pi} \frac{1-r^2}{1-2r\cos(\theta-\phi)+r^2} f(e^{i\phi})\, d\phi, \quad \theta \in [0, 2\pi],$$

is called the convolution of P_r and f. So, the convolution of the Poisson kernel and the boundary value f is the real part of the solution of the Schwarz problem. This idea has been developed to very sophisticated ideas and techniques of singular integral operators and pseudo-differential operators for many problems in analysis and partial differential equations.

Exercises

(1) Let u be a holomorphic function on $\overline{\mathbb{D}}$ such that there exists a nonnegative number M for which

$$|\operatorname{Re} u(z)| \leq M, \quad z \in \partial\mathbb{D}.$$

Prove that for all z in $\mathbb{C}$ with $|z| = r < 1$,

$$|u(z)| \leq |u(0)| + \frac{2M}{1-r}.$$

(2) Let f be a continuous real-valued function on $\partial\mathbb{D}$. Find a harmonic function u on $\mathbb{D}$ such that $u = f$ on $\partial\mathbb{D}$ in the sense that

$$u(re^{i\theta}) \to f(e^{i\theta})$$

uniformly with respect to θ in $[0, 2\pi]$ as $r \to 1-$. (This problem is the famous Dirichlet problem for harmonic functions on $\mathbb{D}$.)

(3) Prove that the solution to the problem in Exercise (2) is unique.

(4) Let f be a continuous function on $\partial\mathbb{D}$. Prove that there is at most one holomorphic function u on $\mathbb{D}$ such that

$$u(re^{i\theta}) \to f(e^{i\theta})$$

uniformly with respect to θ in $[0, 2\pi]$ as $r \to 1-$.

(5) Determine whether or not it is true that for every continuous function f on $\partial\mathbb{D}$, there exists a holomorphic function u on $\mathbb{D}$ such that

$$u(re^{i\theta}) \to f(e^{i\theta})$$

uniformly with respect to θ in $[0, 2\pi]$ as $r \to 1-$. (This problem can be paraphrased as one on whether or not the Dirichlet problem for holomorphic functions on $\mathbb{D}$ is solvable for all continuous boundary values f on $\partial\mathbb{D}$.)

Bibliography

Ahlfors, L. V. (1938). An extension of Schwarz's lemma, *Transactions of the American Mathematical Society* **43**, pp. 359–364.

Ahlfors, L. V. (1979). *Complex Analysis*, 3rd edn. (McGraw-Hill).

Andrews, G. E. (1998). The geometric series in calculus, *American Mathematical Monthly* **105**, pp. 36–40.

Boas, R. P. (1987). *Invitation to Complex Analysis* (Random House).

Burkill, J. C. and Burkill, H. (1980). *A Second Course in Mathematical Analysis. Reprint of the 1970 Original* (Cambridge University Press).

Churchill, R. V. and Brown, J. W. (1990). *Complex Variables and Applications*, 5th edn. (McGraw-Hill)

Copson, E. T. (1944). *An Introduction to the Theory of Functions of a Complex Variable*, 2nd edn. (Clarendon Press).

Copson, E. T. (1975). *Partial Differential Equations* (Cambridge University Press).

Daubechies, I. (1992). *Ten Lectures on Wavelets* (Society for Industrial and Applied Mathematics).

Fefferman, C. (1967). An easy proof of the fundamental theorem of algebra, *American Mathematical Monthly* **74**, pp. 854–855.

Greene, R. E. and Krantz, S. G. (2006). *Function Theory of One Complex Variable*, 3rd edn. (American Mathematical Society).

Hardy, G. H. (1992). *A Course of Pure Mathematics. Reprint of the (1952) Tenth Edition* (Cambridge University Press).

Hardy, G. H. and Rogosinski, W. W. (1999). *Fourier Series* (Dover).

Herstein, I. N. (1975). *Topics in Algebra*, 2nd edn. (Xerox)

Hille, E. (1982). *Analytic Function Theory*, vol. I, 2nd edn. (AMS Chelsea Publishing).

Hille, E. (2005). *Analytic Function Theory*, vol. II (AMS Chelsea Publishing).

Howie, J. M. (2003). *Complex Analysis* (Springer-Verlag).

Knapp, A. W. (1986). *Representation Theory of Semisimple Groups: An Overview Based on Examples* (Princeton University Press).

Krantz, S. G. (2004). *Complex Analysis: The Geometric Viewpoint*, 2nd edn. (The Mathematical Association of America).

Lang, S. (1985). $SL_2(\mathbb{R})$. *Reprint of the 1975 Edition* (Springer-Verlag).
Lang, S. (1990). *Undergraduate Algebra*, 2nd edn. (Springer-Verlag).
Lang, S. (1999). *Math Talks for Undergraduates* (Springer-Verlag).
Lang, S. (1999). *Complex Analysis*, 4th edn. (Springer-Verlag).
Levinson, N. and Redheffer, R. M. (1970). *Complex Variables* (Holden-Day).
Littlewood, J. E. (1944). *Lectures on the Theory of Functions* (Oxford University Press).
Muskhelishvili, N. I. (1977). *Singular Integral Equations* (Noordhott International Publishing).
Nevanlinna, R. and Paatero, V. (1969). *Introduction to Complex Analysis* (Addison-Wesley).
Rudin, W. (1976). *Principles of Mathematical Analysis*, 3rd edn. (McGraw-Hill).
Rudin, W. (1987). *Real and Complex Analysis*, 3rd edn. (McGraw-Hill).
Saff, E. B. and Snider, A. D. (2003). *Fundamentals of Complex Analysis with Applications to Engineering and Science*, 3rd. edn. (Prentice Hall).
Seeley, R. T. (2006). *An Introduction to Fourier Series and Integrals* (Dover).
Stein, E. M. and Shakarchi, R. (2003). *Complex Analysis* (Princeton University Press).
Titchmarsh, E. C. (1923). A contribution to the theory of Fourier transforms, *Proceedings of the London Mathematical Society* **23**, 279–289.
Titchmarsh, E. C. (1939). *The Theory of Functions*, 2nd edn. (Oxford University Press).
Titchmarsh, E. C. (1986). *Introduction to the Theory of Fourier Integrals*, 3rd edn. (Chelsea)
Whittaker, E. T. and Watson, G. N. (1927). *A Course of Modern Analysis*, 4th edn. (Cambridge University Press).
Wong, M. W. (1999). *An Introduction to Pseudo-Differential Operators*, 2nd edn. (World Scientific).
Wong, M. W. (2002). *Wavelet Transforms and Localization Operators*, Operator Theory: Advances and Applications, vol. 136 (Birkhäuser).
Wong, M. W. (2003). Localization operators on the affine group and paracommutators, in H. G. W. Begehr, R. P. Gilbert and M. W. Wong (eds.), *Progress in Analysis*, vol. I (World Scientific), pp. 663–669.
Wong, M. W. (2004). Localization operators, Wigner transforms and paraproducts, in G. A. Barsegian and H. G. W. Begehr (eds.), *Topics in Analysis and its Applications*, NATO Science Series, II. Mathematics, Physics and Chemistry, vol. 147 (Kluwer Academic Publishers), pp. 333–346.
Zygmund, A. (1976). Notes on the history of Fourier series, in J. M. Ash (ed.) *Studies in Harmonic Analysis*, MAA Studies in Mathematics, vol. 13 (The Mathematical Association of America).
Zygmund, A. (2002). *Trigonometric Series* vol. I, II, 3rd edn. (Cambridge University Press).

Index